Studies in Computational Intelligence

Volume 910

The series "Studies in Computational Intelligence" (SCI) publishes new developments and advances in the various areas of computational intelligence—quickly and with a high quality. The intent is to cover the theory, applications, and design methods of computational intelligence, as embedded in the fields of engineering, computer science, physics and life sciences, as well as the methodologies behind them. The series contains monographs, lecture notes and edited volumes in computational intelligence spanning the areas of neural networks, connectionist systems, genetic algorithms, evolutionary computation, artificial intelligence, cellular automata, self-organizing systems, soft computing, fuzzy systems, and hybrid intelligent systems. Of particular value to both the contributors and the readership are the short publication timeframe and the world-wide distribution, which enable both wide and rapid dissemination of research output.

The books of this series are submitted to indexing to Web of Science, EI-Compendex, DBLP, SCOPUS, Google Scholar and Springerlink.

More information about this series at http://www.springer.com/series/7092

Tiansi Dong

A Geometric Approach to the Unification of Symbolic Structures and Neural Networks

Springer

Tiansi Dong
ML2R Competence Center for Machine
Learning Rhine-Ruhr, MLAI Lab, AI
Foundations Group, Bonn-Aachen
International Center for Information
Technology (b-it)
University of Bonn
Bonn, Germany

ISSN 1860-949X ISSN 1860-9503 (electronic)
Studies in Computational Intelligence
ISBN 978-3-030-56274-8 ISBN 978-3-030-56275-5 (eBook)
https://doi.org/10.1007/978-3-030-56275-5

This Springer imprint is published by the registered company Springer Nature Switzerland AG
The registered company address is: Gewerbestrasse 11, 6330 Cham, Switzerland

The Sixth Law of Cognition: Spatial thinking is the foundation of abstract thought.

—Barbara Tversky (2019) p. 72, p. 142

to Elias, Sophia, Peiling, and my parents

Preface

It takes decades to edit a dictionary, to explain each word in terms of other words, and with examples. Alternatively, the neural approach simply learns a vector for each word from texts. We may encounter a new word that is not in our dictionary, but with the neural approach, we can always have a vector of this new word, search its neighbours in vector space, and guess the meaning (without guarantee). Neural-networks (Deep learning) are robust to noisy inputs, able to learn at a level of approximation from enough high-qualified data, but lack of explainability. In contrast, symbolic systems have a set of manually designed rules. This makes outputs explainable and the results guaranteed, but inputs are not robust to noisy inputs. Do the two kinds of systems talk about the same thing ("human intelligence")? Neural people hope this, struggle for decades to design elegant neural-networks that can reach symbolic levels of reasoning, and land at a certain level of approximation. Would they change next year? To answer this question, we designed a very simple experiment. Given a manually edited dictionary that only has tree structured category relation among words, and given vectors of these words provided by well-designed neural-networks, what kinds of mechanism can let these vectors precisely encode the given tree structure? The mechanism that we found is to blow these vectors into balloons in higher dimensional space, so that spatial inclusion relations among balloons could precisely encode tree structured category relations. For symbolists, the configuration of these balloons is a spatial semantics of the tree structure. It seems hard for neural people to change next year, if they restrict vectors as processing objects of neural-networks and fix the dimension of outputs. In a survey of the neural-symbol debates in the literature, we found that the mechanism that we found is an open chapter to solve many questions raised in the debates. Without reservation, we collect these observations in this book, hoping that this could activate symbolists and promote the joint research with Deep learning researchers.

Bonn, Germany
June 2020

Tiansi Dong

Preface

Acknowledgements

First and foremost, I would like to thank Christian Bauckhage, to whom this work owes tremendously. From his reading group at Fraunhofer IAIS to the BMBF P3ML (01/S17064) and ML2R (01/S18038C) projects as well as his encouragement of my leadership role at IPEC (International Program of Excellence) on AI language lab, Christian is a great advocate of my work. I would like to express my deep appreciation to Armin B. Cremers (ABC), who hosted me as a guest scientist at B-IT in 2012 and urged me to pursue a Habilitation, and continues to be an inspiration. ABC has always challenged me to reach for the impossible. I am also indebted to Sören Auer for generously giving me a chance to work with EU H2020 OpenBudgets.eu project in September 2015. A big thank-you to many colleagues, Thomas Bode, Olaf H. Cremers, Martina Dölp, Thomas Fuchs, Bogdan Georgiev, Günter Kniesel, Peter Lachart, Jens Lehmann, Maria Maleshkova, Marie-Luise Liebegut, Sifa Rafet, Alexandra Reitelmann, Daniel Speicher, Thomas Thiel, Susan Tietz, Steffen Wrobel, and Jörg Zimmermann.

For the support of my early work in Spatial Calculus and of organizing Sino-German Symposium GZ1607 "Integrating Symbolic Representation with Numeric Representation for Commonsense Reasoning", I am greatly indebted to Xiaoxing Ma (Nanjing University), and Juanzi Li (Tsinghua University), who also encourages me to challenge hard tasks, and generously shares her research experiences. I am also indebted to Achim Rettinger (University of Trier), Alexander Mehler (Goethe-University of Frankfurt am Main), Barbara Tversky (Stanford and Columbia Universities), Bo Zhang (Tsinghua University), Christian F. Hempelmann (Texas A&M University), Erhard Hinrichs (University of Tübingen), Jie Tang (Tsinghua University), Jun Zhao (Chinese Academy of Sciences), Ron Sun (RPI), Steffen Staab (University of Stuttgart), Xiaoming Fu (Georg-August-University of Göttingen), and Volker Tresp (LMU) for their fruitful discussions during and after the Symposium.

I am also very thankful to Yixin Yang (Northwestern Polytechnical University) for the invitation to Xi'an and the delicious Chinese meals and the delightful conversations in 2018, and to Charles Dee, Xiaoru Meng, Jun Pang, Junfeng Qin, Jianqiu Xu, and Cham Zhong for their friendship.

I am indebted to my parents for their sacrifice and for the happiness they bring whenever they come to visit from China, to members of EFG (Evangelische freie Gemeinde) Bonn, in particular, Margit Jehle, Matthias Jehle, Heidi Raschke, Ebi Raschke, Blanckarts Jürgen, and Ulf Beiderbeck, also to Shih-Kung Cheng and Rui Chen, for the joyful company.

Most importantly, I thank Peiling Cui, for her love and dedicated support and understanding. As a humor researcher, she shows me a third way of facing failure and frustration. Finally, my love to Sophia and Elias for our fun times together.

Bonn, Germany
January 2020

Tiansi Dong

Contents

List of Figures

List of Tables

Chapter 1
Introduction

1.1 Two Paradigms in Artificial Intelligence

The methodology in the research of Artificial Intelligence (AI) consists of two competing paradigms, namely *symbolic approach* and *connectionist approach*. The symbolic approach is based on symbolic structures and rules, in which thinking is reviewed as symbolic manipulation. Associated with this paradigm are features such as *logical, serial, discrete, localized, left-brained*. The connectionist approach is inspired by the physiology of the mind, in which thinking is reviewed as information fusion and transfer of a large network of neurons. Associated with this paradigm are features such as *analogical, parallel, continuous, distributed, right-brained* (Minsky and Papert 1988; Blank et al. 1992).

1.1.1 Symbolic Approach

The symbolic approach has two profound roots: one is the philosophical study of logic, e.g., (Hilbert and Ackermann 1938; Tarski 1946), the other is the study of linguistics, e.g., (Chomsky 1955a, b, 1965).

In elementary school, math teachers would explain the meaning of $1 + 1 = 2$ with concrete examples, such as *one apple plus one apple is two apples, one cat plus one cat is two cats*, …If there were neither $apples$ nor $cats$ in the world, the relation $1 + 1 = 2$ still holds. This relation holds, even if all objects in the world disappear. Pure logic seeks to find the eternal truth in the universe, even concrete worlds disappear (Russell 1919). Influenced by logic, symbolic AI views *intelligence* as logical reasoning, *thought* as rules, and *computation* as symbol transformation. Typically, McCarthy (1995) advocated formal logic to express rules in a pure logical way, so that the completeness and the soundness of solutions could be strictly guaranteed.

T. Dong, *A Geometric Approach to the Unification of Symbolic Structures and Neural Networks*, Studies in Computational Intelligence 910,
https://doi.org/10.1007/978-3-030-56275-5_1

Chomsky (1959) argued that behaviorism is inadequate to account for the ability of learning and using languages. Because any language has an infinite number of syntactically well-formed sentences, speakers can understand and produce sentences that they had not previously encountered. Chomsky developed generative grammar in terms of automata that could generate sentences.

Symbolic approaches deal with (1) the representation and construction of symbols, (2) programs that manipulate these symbols, and (3) descriptions of intelligent behaviors. A typical symbolic system consists of three components as follows (Dinsmore 1992):

- symbols: some are primitives, others are constructed using primitive symbols;
- combinatorial semantics: the meaning of constructed symbols can be interpreted by the meaning of primitive symbols and the way of the construction;
- reasoning, which is conducted as symbol manipulation.

A symbol must symbolize something physical or conceptual. The process of problem solving is viewed as the state transition from the initial state to an end state. States and transition rules are expressed in symbolic forms. Newell and Simon (1976) proposed *Physical Symbol System Hypothesis* (PSSH) that *a physical symbol system has the necessary and sufficient means for general intelligence action.*

1.1.2 Connectionist Approach

Different from symbolic approach, the connectionist approach was initially inspired by the observation that the functioning of the mind is a network of neurons. The connectionist approach views intelligence as activities of interconnected neurons.

Connectionists, from the very beginning, have been aiming at understanding how the mind works through the simulation of networks, e.g., how memories are established? how associations are stored for patterns? McCulloch and Pitts (1943) proposed that a simple connectionist model can perform logical computation, such as $\wedge$, $\vee$, $\neg$. These mathematical neurons are two-valued nodes, representing either *on* or *off*. Hebb (1949) suggested that when two neurons were jointly active, the strength of the connection might be increased. The approach can be understood as embodying the biological mechanism of human intelligence by connectionist networks. Rosenblatt (1962) used connectionist networks for pattern recognition, a nightmare task for symbolists. He researched statistical patterns over multiple units, and treated noise and variation as essential. Selfridge (1988) explored the recognition of letters, a task being difficult due to the fact that different people write their letters differently.

In terms of networks, rather than of symbols, connectionists have achieved a new type of information processing system that is carried out by exciting or inhibiting nodes. Knowledge is represented as weights of connections between nodes. The problem solving is a process of gearing weights to establish a harmony between input and output in the training data. In the task of object recognition, a connectionist will represent input images and a set of candidate object classes as vectors. The ith element in the output vector represents the ith object class. The whole network will be geared (updating weights) by a large number of images and their known classes. After that, the output from an input of a new image is regarded as the predicting behavior of this network. In this way, connectionist networks approximate subsymbolic features of entities, which are below symbolic manipulations of entities. Smolensky (1988) proposed *Subsymbolic Hypothesis* (SH) that *the intuitive processor is a sub-conceptual connectionist dynamical system that does not admit a complete, formal, and precise conceptual level description*.

1.1.3 Complementary Strength

The two approaches have complementary strength—The symbolic approach is good at reasoning, weak in learning, vulnerable to noisy or unwanted inputs. Sun (2015) listed a number of limitations of symbolic approaches, for example, symbolic representations are fixed, determined a priori, hand-made, therefore, could become extremely expensive in real applications. The connectionist approach is good at learning by utilizing gradient descent methods, e.g., back propagation (Rumelhart et al. 1986, 1988), robust for noisy inputs, and suffer from three limitations as follows:

- It is hard to develop connectionist knowledge representations;
- Developed connectionist knowledge representations are lack of explainability;
- As reasoning is carried out by minimizing error functions, connectionist reasoning is limited at approximation.

Remedies to the weakness of the symbolic approach include Fuzzy logic and statistical approaches. Zadeh (1965) proposed fuzzy logic, in which truth value is understood as a degree. The truth value is extended from two discrete values to a continuum from 0 to 1. He coined the membership function to measure the degree with which an object belongs to a set. In this way, fuzzy logic can be used to represent vagueness and approximate information, such as "warm", "tall". However, Elkan (1993) found that fuzzy logic will be degraded into classic logic under certain unintended conditions.

Pearl (1988) used Bayesian approach as a probabilistic remedy to the brittleness of symbolic approaches. However, humans do not always follow probabilistic rules to think. In reality, obtaining adequate data for statistical analysis may not be easy.

The connectionist approach, in particular the Deep Learning approach, has regained popularity (LeCun et al. 2015; Goodfellow et al. 2016), and has demonstrated great performances in a variety of AI tasks, such as image recognition (Krizhevsky et al. 2012; Farabet et al. 2013; Tompson et al. 2014; Szegedy et al. 2014), speech recognition (Mikolov et al. 2011; Hinton et al. 2012a; Sainath et al. 2015), natural language processing (Collobert et al. 2011; Mikolov et al. 2013; Bordes et al. 2014; Sutskever et al. 2014; Jean et al. 2015), bioinformatics (Leung et al. 2014; Xiong et al. 2015; Ma et al. 2015), particle analysis in chemistry (Ciodaro et al. 2012;

Helmstaedter et al. 2013). The most striking achievement of Deep Learning would be AlphaGo (Silver et al. 2017) that is the first AI system to defeat Go world champion.

However, recent studies found that Deep Learning Systems can be easily fooled by adversarial samples (Seo et al. 2016; Jia and Liang 2017; Belinkov and Bisk 2017), and that they normally need much more learning data than human does (Lake et al. 2015), and are unfortunately not absolutely robust for safety critical applications, such as autonomous driving.

1.2 Recent Debates on Structure and Learning

Intelligence can be simulated neither as truth-false reasoning, nor as back-propagation learning processing alone. It would be better a synergistic integration. A discussion on imposing innate structural constraints into Deep Learning systems was held at Stanford AI salon in February 2018. One party (Christopher Manning) argued for the necessity of imposing innate structures, the counterpart was Yann LeCun. Both believed that structure is necessary. Manning views structure as necessary good, because it helps a system to learn from less amount of data, to be able to learn at higher level of abstractions. LeCun argues structure as necessary evil, as it can be invariably misled by some data, or become obsolete in the future. After observing that human babies are able to learn about world through observation without external reward, and to learn concepts without supervision, they questioned whether it is necessary to impose structure first. Manning argued that we should provide some structures to make learning more efficiently. LeCun believes it possible to learn all structures from the environment. The debate would have been more concrete, if before the debate, there had been a Deep Learning system precisely imposed by some structures (we will revisit part of the debate in Sect. 8.2.1).

1.3 Logic and Its Two Perspectives

Logic is a formal language that consists of *terms* and *formulas* (Tarski 1946; Blackburn et al. 2001). In first order logic, *terms* can be constants, variables, and functions; *formulas*, either a simple relation, or built from boolean operators and quantifiers, are assertions on a structure of *terms*. For example, if 1 and 2 are constant *terms*, and $+$ is a function *term*, then $1 + 2$ is also a *term*; $2 > 1$, $1 + 1 = 2$ are simple relation *formulae*; $2 > 1 \wedge 1 + 1 = 2$ is *formula* built from boolean relation, $\exists x[x + 1 > 2]$ is *formula* built from quantifier. Logical deduction, namely *modus ponens*, take the form as follows.

$$P(u), \forall x[P(x) \to Q(x)] \vdash Q(u) \tag{1.1}$$

read as *$P(u)$ is true; if for any x, $P(x)$ is true, then $Q(x)$ is true; $Q(u)$ is true*, which can be represented spatially by Venn diagram (Venn 1880). The parallel reading for

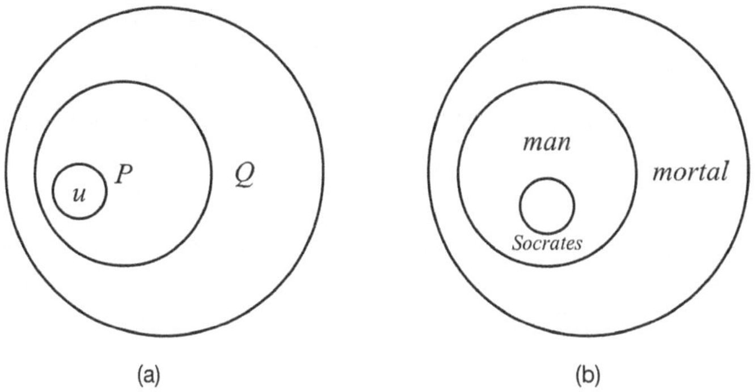

Fig. 1.1 **a** Venn diagram for the deduction that *(1) $P(u)$ is true; (2) for every x, if $P(x)$ is true, then $Q(x)$ is true; therefore, $Q(u)$ is true*; **b** Venn diagram for the assertion that *if all men are mortal and Socrates is a man, then Socrates is mortal*

$\forall x[P(x) \to Q(x)]$ would be *if for any region x, x is inside Region P, then x is inside Region Q*. The *modus ponens* rule is spatialized by the relations between regions.

Region u is located inside region P that is inside region Q, therefore, Region u is inside Region Q.

A well-known deduction example is as follows: given *all men are mortal; Socrates is a man*, we can deduce *Socrates is mortal*. In terms of Venn diagram, this deduction can be diagrammatically represented as follows: if *Socrates* region is inside *man* region that is inside *mortal* region, *Socrates* region is thus inside *mortal* region, as illustrated in Fig. 1.1.

Here, we illustrate two perspectives of logic: the proof perspective and the model perspective. The proof perspective focuses on the forms of logic (the syntax), and is targeted at truth-preserving of propositions. A proof is a sequence of propositions, each is either an axiom, or a theorem derived by applying deduction rules for propositions that have appeared in earlier positions in the sequence. The model theoretic perspective views symbolic propositions as denoting devices—a proposition is true, if it denotes something true in the world. That is, propositions are viewed as descriptions about a model—a proposition will be true, if its denotation, e.g., an entity, a feature of an entity, or a relation, exists in a model. The magic of logic is the relation between the two perspectives: to check whether a new assertion has a model, we can focus on a piece of pen-paper work to construct a proof based on very limited number of axioms and deduction rules, instead of searching a possible model in the endless universe. If we syntactically proof it true, we know this new assertion has a model. This proof perspective dominates the paradigm of symbolic approach, as well as the hybrid approaches (see Sect. 2.5).

Connectionist networks, including Deep Learning, are capable of capturing association relations in huge amount of training data. For example, words that co-occur in contexts, frames that co-occur in videos, or paired-corpus for machine translation, images with their labels. Connectionist approaches assume that things that occur in similar contexts or environments tend to have some similar features (Wittgenstein 1953; Harris 1954), and represent things in terms of vectors, so that the similarity can be numerically measured (for example, by the cos value between two vectors). These vectors are the model constructed by connectionist networks. Intuitively, integrating connectionist and symbolic models can be achieved by extending vectors, produced by connectionist network, into regions under the condition that spatial relations among regions capture symbolic structures, as Venn diagram does. We propose geometric construction method to realize such kind of integration. That is, we spatialize symbolic structures onto a vector space. So comes the name *Geometric Connectionist Machine for symbol spatialization.*

1.4 Geometric Connectionist Machine

> It has seemed to me for some years now that there must be a unified account in which the so-called rule-governed and exceptional cases were dealt with by a unified underlying process—a process which produces rule-like and rule-exception behavior through the application of a single process ...[In this process] ...both the rule-like and non-rule-like behavior is a product of the interaction of a very large number of 'sub-symbolic' processes.
>
> (Rumelhart 1984, p.60)

Natural language processing is a favorite topic of symbolic approaches. A basic symbolic structure of words is the taxonomy of hypernym relations among word-senses. For example, in WordNet 3.0 (Miller 1995), there are hypernym chains such as *national_capital.n.01* $\subset$ *city.n.01* $\subset$ *municipality.n.01* $\subset$ *urban_area.n.01*. Here, the syntax structure takes the form of

$$\text{word-sense 1} \subset \text{ord-sense 2} \tag{1.2}$$

The rule of reasoning is limited to the transitive relation as follows.

$$x \subset y \wedge y \subset z \rightarrow x \subset z \tag{1.3}$$

A spatial model that captures this structure is regions and inclusion relations among regions.

1.4.1 An Approach of Geometric Construction

We show a geometric construction method that is able to precisely transform tree structured hypernym relations of word-senses onto $\mathcal{N}$-Balls (balls in super-dimensional space) (Dong et al. 2019b, c). These $\mathcal{N}$-Balls have features as follows:

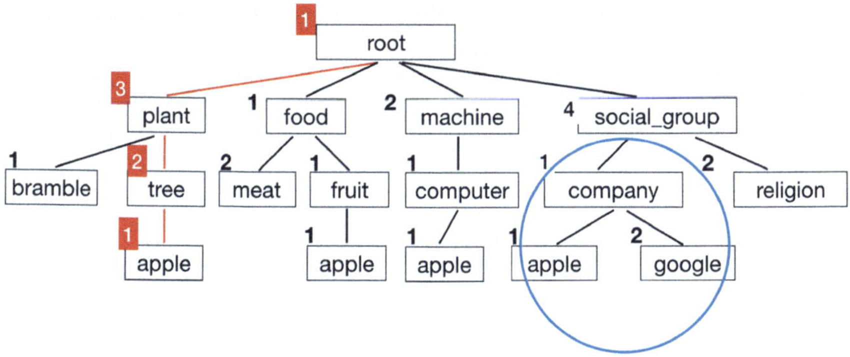

Fig. 1.2 Four taxonomy structures of apple are merged into one tree. The path to apple, as a kind of tree, from root is uniquely identified as the path-vector (Minsky and Papert 1988; Blank et al. 1992; Hilbert and Ackermann 1938)

- The dimension of the $\mathcal{N}$-Ball of a word-sense is higher than that of vector representation of the word stem;
- $\mathcal{N}$-Balls preserve the vector representation of the word stem very well;
- Symbolic tree structures of hypernym relations are precisely (at zero energy cost) represented by inclusion relations among $\mathcal{N}$-Balls.

Given a number of tree structures, we merge them into a big tree structure, and introduce a common root, if needed. Sibling nodes are alphabetically ordered, each with an ordinal number that creates a unique path from the root, as illustrated in Fig. 1.2. Each node in the big tree structure shall be represented by an $\mathcal{N}$-Ball whose radius precisely satisfies the inclusion condition: *A is a child of B, if and only if A's $\mathcal{N}$-Ball is contained by B's $\mathcal{N}$-Ball*. The central point of the $\mathcal{N}$-Ball has an orientation fixated by concatenating three vectors: (1) the vector representation of the word stem produced by a Deep Learning system, (2) the fixed-length path-vector from the root to its parent node in the big tree structure, (3) a constant vector which avoids the $\mathcal{N}$-Ball containing the origin point of the space in the later geometrical transformation process.

We adopt depth-first recursive process to traverse the nodes in the tree structure, meanwhile constructing $\mathcal{N}$-Balls through three geometric transformations: (1) homothetic transformation, which zooms the radius and the length of the central point with the same ratio, (2) shift transformation, which adds a vector to the central point, and (3) rotation transformation, which rotates the orientation of the central point. To prevent the deterioration of already improved relations, we will apply the same transformation for all its child balls, when we apply a geometric transformation for an $\mathcal{N}$-Ball (see Sects. 5.4 and 5.5).

Every $\mathcal{N}$-Ball is initialized with the same radius and the same length of the central vector point, as illustrated in Fig. 1.3a. To guarantee the disconnectedness relation among sibling $\mathcal{N}$-Balls, we use homothetic transformation to push connected sibling

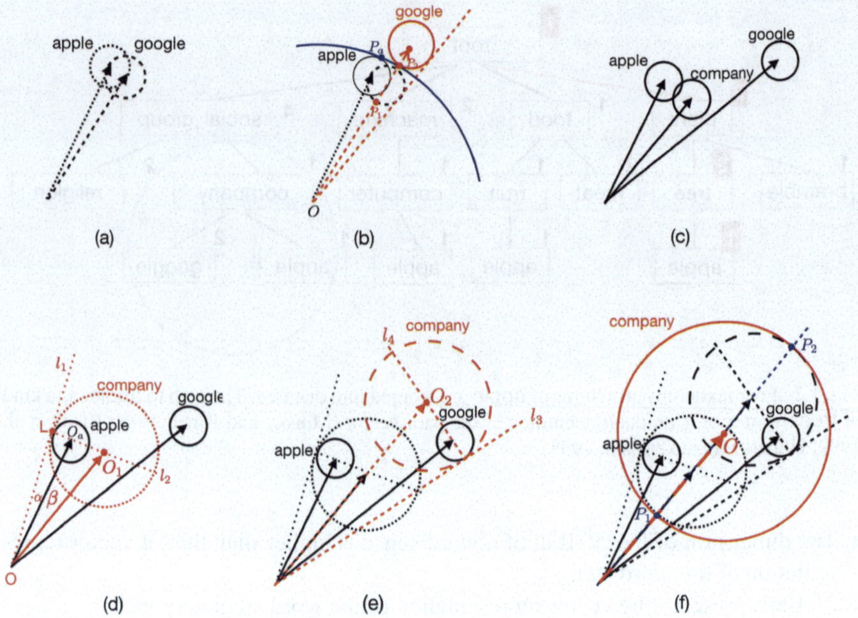

Fig. 1.3 Construct $\mathcal{N}$-Ball embeddings of a simple tree structure of apple, google, company in the blue circle in Fig. 1.2

$\mathcal{N}$-Balls farther away from each other. If the initial google $\mathcal{N}$-Ball connects with apple $\mathcal{N}$-Ball, we will apply a homothetic transformation for the google ball. The ratio of the homothetic transformation is $\frac{|OP_a|}{|OP_1|}$, in which P_a is the apogee of apple's $\mathcal{N}$-Ball, P_1 is the perigee of google's $\mathcal{N}$-Ball, as illustrated in Fig. 1.3b, P_2 is the new perigee after transformation satisfying $|OP_a| = |OP_2|$, that is, the remote distance of apple's $\mathcal{N}$-Ball equals to the near ground distance of google's $\mathcal{N}$-Ball.

Following the depth-first procedure, we construct the parent $\mathcal{N}$-Ball after all child $\mathcal{N}$-Balls are constructed. That is, company $\mathcal{N}$-Ball will be initialized, after apple $\mathcal{N}$-Ball and google $\mathcal{N}$-Ball are settled down (Fig. 1.3c). Then, for each child $\mathcal{N}$-Ball, we construct an $\mathcal{N}$-Ball under two conditions: (1) its central point vector is along the central point vector of this initialized company $\mathcal{N}$-Ball; (2) it properly contains this child $\mathcal{N}$-Ball. As illustrated in Fig. 1.3d, we construct a company $\mathcal{N}$-Ball that properly contains apple $\mathcal{N}$-Ball. In the same way, we construct a company $\mathcal{N}$-Ball that properly contains google $\mathcal{N}$-Ball, as illustrated in Fig. 1.3e. The final company $\mathcal{N}$-Ball is the minimal cover of all created company $\mathcal{N}$-Balls, as illustrated in Fig. 1.3f.

1.4.2 *Two Kinds of Geometric Connectionist Machines*

We have created two kinds of Geometric Connectionist Machines: $\mathbf{GCM_0}$ and $\mathbf{GCM_1}$. $\mathbf{GCM_0}$ is for spatializing unlabeled symbolic tree structures onto the vector space, and has been experimented by spatializing hypernym trees of word-senses extracted from WordNet (Miller 1995) onto GloVe word-embeddings (Pennington et al. 2014). Precise spatialization has been achieved. Compared with pre-trained GloVe embeddings, $\mathcal{N}$-Ball embeddings produce clean semantics in the nearest neighborhood experiment, and overcome the data sparse problem. For example, *tiger* as an audacious person (tiger.n.01) and *linguist* as a specialist in linguistics (linguist.n.02) seldom appear in the same context, their GloVe embeddings produce -0.1 as the cosine similarity. However, they are hyponyms of person.n.01. Using this structural information, our geometrical process transform the $\mathcal{N}$-Balls of tiger.n.01 and linguist.n.02 inside the $\mathcal{N}$-Ball of person.n.01, resulting in high similarity value in their $\mathcal{N}$-Ball embeddings. In our imposed structure, as listed in Table 1.1, 'france.n.02' (as a family name of authors), 'journalist.n.01' (as a pro-

Table 1.1 Top-6 $\mathcal{N}$-Ball nearest neighbors compared with top-N GloVe nearest neighbors. $\mathcal{N}$-Ball neighbors possess clean semantics

word-sense 1	word-sense 2	word
beijing.n.01	london.n.01, atlanta.n.01	China, Taiwan, Seoul, Taipei, Chinese, Shanghai
	washington.n.01, paris.n.0	Korea, Mainland, Hong, Wen, Kong, Japan
	potomac.n.02, boston.n.01	Hu, Guangzhou, Chen, visit, here, Tokyo, Vietnam
berlin.n.01	madrid.n.01, toronto.n.01	Vienna, Warsaw, Munich, Prague, Germany
	rome.n.01, columbia.n.03	Moscow, Hamburg, Bonn, Copenhagen, Cologne
	sydney.n.01, dallas.n.01	Dresden, Leipzig, Budapest, Stockholm, Paris
berlin.n.02	imon.n.02, williams.n.01	Frankfurt, Amsterdam, German, Stuttgart, Brussels
	foster.n.01, dylan.n.01	Petersburg, Rome, Austria, Bucharest, Düsseldorf
	mccartney.n.01, lennon.n.01	Zurich, Kiev, Austrian, Heidelberg, London
tiger.n.01	survivor.n.02, neighbor.n.01	Tigers, Woods, Warrior, Ltte, Wild, Elephant
	immune.n.01, linguist.n.02	Crocodile, Leopard, Eelam, Warriors, Elephants
	bilingual.n.01, warrior.n.01	Eagle, Hunt, Dog, Jungle, Lone, Cat, Hunting
france.n.02	white.n.07, woollcott.n.01	French, Belgium, Paris, Spain, Netherlands
	uhland.n.01, london.n.02	Italy, Germany, European, Switzerland, Europe
	journalist.n.01, poet.n.01	Belgian, Dutch, Britain, Portugal, Luxembourg

fession), 'poet.n.01' (as a profession) share the same direct hypernym 'writer.n.01'. In contrast, GloVe neighbors are biased by the training corpus, mixed with unintended words (in red color). For example, neighbors of 'berlin' neglect an important word-sense of 'berlin' as a family name.

$\mathbf{GCM}_1$ is an extension from $\mathbf{GCM}_0$ for spatializing labeled tree structures extracted from knowledge graphs, and applied for Triple Classification task. In Knowledge Graph, Triples have the form of (*subject, predicate, object*). We can view a Triple as a three-word sentence, and learn vector embeddings using connectionist networks (Bordes et al. 2013). On the other hand, knowledge-graph has symbolic structures. Given subject s and predicate p, we can find all objects $o_1, \ldots, o_n$ satisfying $(s, p, o_i), i = 1, \ldots, n$, and a hypernym chain of s: $s_0(= s), s_1, s_2, \ldots, s_k$. For example, *city* has members *new_york, london, shanghai, berlin*, and a hypernym chain: *city*, *municipality*, *region*, *physical_entity*, *entity*. This chain and all Triples (s, p, o_i)s form a simple labeled tree structure. We spatialize this labeled tree onto the vector embeddings of their nodes, and record the transformation sequence of each $\mathcal{N}$-Ball. To predict whether a new entity x is an object of subject s_0 with relation p, we assume that (s_0, p, x) stands, and apply the geometric process of o_1 to x. If the $\mathcal{N}$-Ball of x is located inside the same $\mathcal{N}$-Ball as o_1, (s_0, p, x) will be predicted as true. This method can be likened to way-finding in everyday life (see Sect. 6.5.4 for a detailed description). The results of our experiments show that our approach greatly outperforms traditional knowledge-embedding approaches, especially when the type chain is long (Dong et al. 2019c).

1.5 Towards the Two-System Model of the Mind

Connectionists' efforts to achieve symbolic reasoning can date back to Hinton (1981), and still active (Dong et al. 2019a). Hinton (1981) applied connectionist approaches for reasoning with knowledge graphs, as vector embedding is an important method to explain the *similarity judgment* of Tversky (1977) and to simulate one of three judgment methods under uncertainty (Tversky and Kahneman 1974), whose thoughts are later further developed and summarized by Kahneman (2011) in terms of *two systems of the mind*: one system for associative memory (System 1), the other for logic reasoning (System 2).

One function of the mind (System 1) coheres a causally linked story (or model), thus representing the normal understanding of the given input data. Connectionist networks somehow simulate System 1 by constructing a harmony between datasets (inputs) and their labels (outputs) (supervised machine learning) (Socher et al. 2011; Hinton et al. 2012b; Vukotic et al. 2017), or among datasets themselves (unsupervised machine learning) (Goodfellow et al. 2016). The second function of the mind (System 2) is the logical reasoning that follows rules, calculates, and reasons, and is in charge of doubt and disbelieving (Kahneman 2011).

The two systems synergistically work together. Only with System 1, we would believe almost anything (Kahneman 2011, p.81). We propose a geometrical method

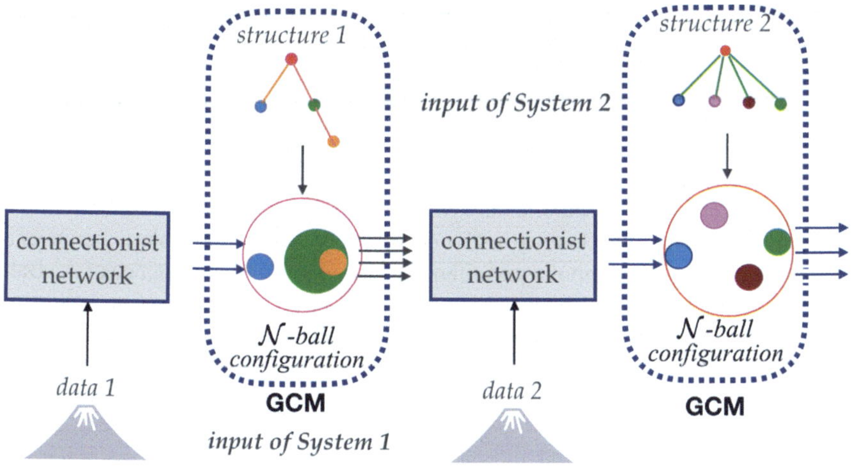

Fig. 1.4 An architecture of integrating connectionism with symbolism through symbol spatialization

to spatialize representations of System 2 onto representations of System 1. In this way, we shall have an aggregated model that is more robust than Deep Learning system alone. For a precise spatialization, we have to abandon the back-propagation method, and adopt geometric construction processes to create $\mathcal{N}$-Ball configuration. As an $\mathcal{N}$-Ball can be represented by a vector of its central point and the radius, the spatialization process can be done repeatedly, as shown in Fig. 1.4. The synergistic interaction between System 1 and System 2 provides valuable cues for creating a novel synergistic hybrid model that is on the way to fill the gap between symbolic and connectionist approaches.

The rest of the book is structured as follows:

- Chapter 2 is on the gap between symbolic and connectionist approaches. We review different and common features between the two approaches, and also the philosophies for closing the gap. We visit classic connectionist's attempts to bridge the gap, and describe the problem of hybrid approaches;
- Chapter 3 reviews the two-system model of the mind. We argue for the region-based representation for the integration of connectionist networks and symbolic structures, and formalize the symbol spatialization as a task of constructing Geometric Connectionist Machines;
- Chapter 4 presents the criteria and challenges for symbol spatialization. We revisit the back-propagation method, and shows that it fails to satisfy our criteria;
- Chapter 5 lists design principles of Geometric Connectionist Machines, and presents the algorithm for spatializing an unlabeled tree using geometric construction;

- Chapter 6 develops the first Geometric Connectionist Machine. We apply the geometric approach for the creation of a large scaled $\mathcal{N}$-Ball embeddings, and evaluate the quality;
- Chapter 7 develops the second Geometric Connectionist Machine $\mathbf{GCM}_1$ that accepts labeled tree structures, and applies $\mathbf{GCM}_1$ for the task of Triple Classification in Knowledge Graphs;
- Chapter 8 revisits the symbolic-subsymbolic debates, and shows the $\mathcal{N}$-Ball configurations created by Geometric Connectionist Machines are able to serve as a continuum between symbolic and connectionist models. The existence of $\mathcal{N}$-Ball configurations resolves the antagonism between Connectionism and Symbolicism;
- Chapter 9 concludes the book and lists a number of research topics.

References

Belinkov, Y., & Bisk, Y. (2017). Synthetic and natural noise both break neural machine translation. CoRR arXiv:abs/1711.02173.

Blackburn, P., de Rijke, M., & Venema, Y. (2001). *Modal logic*. New York, NY, USA: Cambridge University Press.

Blank, D. S., Meeden, L. A., & Marshall, J. (1992). Exploring the symbolic/subsymbolic continuum: A case study of RAAM. In *The symbolic and connectionist paradigms: Closing the gap* (pp. 113–148). Erlbaum.

Bordes, A., Chopra, S., & Weston, J. (2014). Question answering with subgraph embeddings. In *Proceedings of the 2014 Conference on Empirical Methods in Natural Language Processing (EMNLP)* (pp. 615–620). Doha, Qatar: Association for Computational Linguistics.

Bordes, A., Usunier, N., Garcia-Duran, A., Weston, J., & Yakhnenko, O. (2013). Translating embeddings for modelling multi-relational data. In C. J. C. Burges, L. Bottou, M. Welling, Z. Ghahramani, & K. Q. Weinberger (Eds.), *Advances in neural information processing systems* (Vol. 26, pp. 2787–2795). Curran Associates, Inc.

Chomsky, N. (1955a). *Logical structure of linguistic theory*. MIT Humanities Library, Microfilm.

Chomsky, N. (1955b). *Syntactic structures*. The Hague: Mouton.

Chomsky, N. (1959). On certain formal properties of grammars. *Information and Control*, *2*(2), 137–167.

Chomsky, N. (1965). *Aspects of the theory of syntax*. Massachusetts: The MIT Press.

Ciodaro, T., Deva, D., de Seixas, J. M., & Damazio, D. (2012). Online particle detection with neural networks based on topological calorimetry information. *Journal of Physics: Conference Series*, *368*, 012030.

Collobert, R., Weston, J., Bottou, L., Karlen, M., Kavukcuoglu, K., & Kuksa, P. (2011). Natural language processing (almost) from scratch. *Journal of Machine Learning Research*, *12*, 2493–2537.

Dinsmore, J. (1992). Thunder in the gap. In *The symbolic and connectionist paradigms: closing the gap* (pp. 1–23). Erlbaum.

Dong, H., Mao, J., Lin, T., Wang, C., Li, L., & Zhou, D. (2019a). Neural logic machines. In *ICLR-19*, New Orleans, USA.

Dong, T., Bauckhage, C., Jin, H., Li, J., Cremers, O. H., Speicher, D., Cremers, A. B., & Zimmermann, J. (2019b). Imposing category trees onto word-embeddings using a geometric construction. In *ICLR-19*, New Orleans, USA, 6-9 May 2019.

Dong, T., Wang, Z., Li, J., Bauckhage, C., & Cremers, A. B. (2019c). Triple classification using regions and fine-grained entity typing. In *Proceedings of the Thirty-Third AAAI Conference on Artificial Intelligence (AAAI-19)* (pp. 77–85), Honolulu, Hawaii, USA. 27 January–1 February 2019.

Elkan, C. (1993). The paradoxical success of fuzzy logic. *IEEE Expert*, 698–703.

Farabet, C., Couprie, C., Najman, L., & LeCun, Y. (2013). Learning hierarchical features for scene labeling. *IEEE Transactions on Pattern Analysis and Machine Intelligence*, *35*(8), 1915–1929.

Goodfellow, I., Bengio, Y., & Courville, A. (2016). *Deep learning*. The MIT Press.

Harris, Z. (1954). Distributional structure. *Word*, *10*(23), 146–162.

Hebb, D. (1949). *The organization of behavior: A neuropsychological theory*. Washington, USA: Psychology Press.

Helmstaedter, M., Briggman, K. L., Turaga, S. C., Jain, V., Seung, H. S., & Denk, W. (2013). Connectomic reconstruction of the inner plexiform layer in the mouse retina. *Nature*, *500*, 168–174.

Hilbert, D., & Ackermann, W. (1938). *Principles of mathematical logic*. Berlin. Citation based on the reprinted version by the American Mathematical Society (1999)

Hinton, G. E. (1981). Implementing semantic networks in parallel hardware. In G. E. Hinton & J. A. Anderson (Eds.), *Parallel models of associative memory* (pp. 161–187). Hillsdale, NJ: Erlbaum.

Hinton, G., Deng, L., Yu, D., Dahl, G., Rahman Mohamed, A., Jaitly, N., Senior, A., Vanhoucke, V., Nguyen, P., Sainath, T., & Kingsbury, B. (2012a). Deep neural networks for acoustic modeling in speech recognition. *Signal Processing Magazine*, *29*(6), 82–97

Hinton, G., Deng, L., Yu, D., Dahl, G. E., Mohamed, A.-R., Jaitly, N., et al. (2012b). Deep neural networks for acoustic modeling in speech recognition: The shared views of four research groups. *IEEE Signal Processing Magazine*, *29*(6), 82–97.

Jean, S., Cho, K., Memisevic, R., & Bengio, Y. (2015). On using very large target vocabulary for neural machine translation. In *Proceedings of the 53rd Annual Meeting of the Association for Computational Linguistics and the 7th International Joint Conference on Natural Language Processing (Volume 1: Long Papers)* (pp. 1–10). Beijing, China: Association for Computational Linguistics.

Jia, R., & Liang, P. (2017). Adversarial examples for evaluating reading comprehension systems. In *Proceedings of the 2017 Conference on Empirical Methods in Natural Language Processing, EMNLP 2017, Copenhagen, Denmark, 9–11 September 2017* (pp. 2021–2031).

Kahneman, D. (2011). *Thinking, fast and slow*. Allen Lane, Penguin Books. Nobel laureate in Economics in 2002.

Krizhevsky, A., Sutskever, I., & Hinton, G. E. (2012). Imagenet classification with deep convolutional neural networks. In *Proceedings of the 25th International Conference on Neural Information Processing Systems—Volume 1*, NIPS'12 (pp. 1097–1105). USA: Curran Associates Inc.

Lake, B., Salakhutdinov, R., & Tenenbaum, J. (2015). Human-level concept learning through probabilistic program induction. *Science*, *350*(6266), 1332–1338.

LeCun, Y., Bengio, Y., & Hinton, G. E. (2015). Deep learning. *Nature*, *521*(7553), 436–444.

Leung, M. K. K., Xiong, H. Y., Lee, L. J., & Frey, B. J. (2014). Deep learning of the tissue-regulated splicing code. *Bioinformatics*, *30*(12), 121–129.

Ma, J., Sheridan, R. P., Liaw, A., Dahl, G. E., & Svetnik, V. (2015). Deep neural nets as a method for quantitative structure activity relationships. *Journal of Chemical Information and Modeling*, *55*(2), 263–274.

McCarthy, J. (1995). Programs with common sense. In G. F. Luger (Ed.), *Computation and intelligence* (pp. 479–492). Menlo Park, CA, USA: American Association for Artificial Intelligence.

McCulloch, W. S., & Pitts, W. (1943). A logical calculus of the ideas immanent in nervous activity. *Bulletin of Mathematical Biophysics*, *5*, 115–133.

Mikolov, T., Deoras, A., Povey, D., Burget, L., & Cernocký, J. (2011). Strategies for training large scale neural network language models. In D. Nahamoo & M. Picheny (Eds.), *ASRU* (pp. 196–201). IEEE.

Mikolov, T., Yih, W.-T., & Zweig, G. (2013). Linguistic regularities in continuous space word representations. *Proceedings of NAACL-HLT,* 746–751.
Miller, G. A. (1995). Wordnet: A lexical database for english. *Communication of ACM, 38*(11), 39–41.
Minsky, M., & Papert, S. (1988). *Perceptrons*. Cambridge, MA, USA: MIT Press.
Newell, A., & Simon, H. A. (1976). Computer science as empirical inquiry: Symbols and search. *Communication of ACM, 19*(3), 113–126.
Pearl, J. (1988). *Probabilistic reasoning in intelligent systems: Networks of plausible inference*. San Francisco, CA, USA: Morgan Kaufmann Publishers Inc.
Pennington, J., Socher, R., & Manning, C. D. (2014). GloVe: global vectors for word representation. In *EMNLP'14* (pp. 1532–1543).
Rosenblatt, F. (1962). *Principles of neurodynamics: Perceptrons and the theory of brain mechanisms*. Washington, USA: Spartan Books.
Rumelhart, D. E. (1984). The emergence of cognitive phenomena from sub-symbolic processes. In *Proceedings of the Sixth Annual Conference of the Cognitive Science Society* (pp. 59–62). Hillsdale, NJ and Bolder, Colorado: Erlbaum.
Rumelhart, D. E., Hinton, G. E., & Williams, R. J. (1988). Neurocomputing: Foundations of research. In *Learning representations by back-propagating errors* (pp. 696–699). MIT Press, Cambridge, MA, USA.
Rumelhart, D. E., McClelland, J. L., & PDP Research Group (Eds.). (1986). *Parallel distributed processing: Explorations in the microstructure of cognition, Vol. 1: Foundations*. MIT Press, Cambridge, MA, USA.
Russell, B. (1919). *Introduction to mathematical philosophy*. George Allen & Unwin, Ltd., London and The Macmillan Co., New York. Citation is based on the reprint by Dover Publications, Inc. (1993).
Sainath, T. N., Kingsbury, B., Saon, G., Soltau, H., Mohamed, A.-R., Dahl, G., & Ramabhadran, B. (2015). Deep convolutional neural networks for large-scale speech tasks. *Neural Network, 64*(C), 39–48.
Selfridge, O. G. (1988). Pandemonium: A paradigm for learning. In J. A. Anderson & E. Rosenfeld (Eds.), *Neurocomputing: Foundations of research* (pp. 115–122). Cambridge, MA, USA: MIT Press.
Seo, M. J., Kembhavi, A., Farhadi, A., & Hajishirzi, H. (2016). Bidirectional attention flow for machine comprehension. CoRR arXiv:abs/1611.01603.
Silver, D., Schrittwieser, J., Simonyan, K., Antonoglou, I., Huang, A., Guez, A., et al. (2017). Mastering the game of go without human knowledge. *Nature, 550*, 354–359.
Smolensky, P. (1988). On the proper treatment of connectionism. *Behavioral and Brain Sciences, 1*, 1–23.
Socher, R., Pennington, J., Huang, E. H., Ng, A. Y., & Manning, C. D. (2011). Semi-supervised recursive autoencoders for predicting sentiment distributions. In *Proceedings of the Conference on Empirical Methods in Natural Language Processing*, EMNLP '11 (pp. 151–161). Stroudsburg, PA, USA: Association for Computational Linguistics.
Sun, R. (2015). Artificial intelligence: Connectionist and symbolic approaches. In D. W. James (Ed.), *International encyclopedia of the social and behavioral sciences* (2nd ed., pp. 35–40). Oxford: Pergamon/Elsevier.
Sutskever, I., Vinyals, O., & Le, Q. V. (2014). Sequence to sequence learning with neural networks. In *Proceedings of the 27th International Conference on Neural Information Processing Systems—Volume 2*, NIPS'14 (pp. 3104–3112). Cambridge, MA, USA: MIT Press.
Szegedy, C., Liu, W., Jia, Y., Sermanet, P., Reed, S. E., Anguelov, D., Erhan, D., Vanhoucke, V., & Rabinovich, A. (2014). Going deeper with convolutions. CoRR arXiv:abs/1409.4842.
Tarski, A. (1946). *Introduction to logic and to the methodology of deductive sciences*. Oxford University Press, New York. Citation based on the Dover edition, first published in 1995.

Tompson, J., Jain, A., LeCun, Y., & Bregler, C. (2014). Joint training of a convolutional network and a graphical model for human pose estimation. In *Proceedings of the 27th International Conference on Neural Information Processing Systems—Volume 1*, NIPS'14 (pp. 1799–1807). Cambridge, MA, USA: MIT Press.

Tversky, A. (1977). Features of similarity. *Psychological review*, *84*, 327–353.

Tversky, A., & Kahneman, D. (1974). Judgment under uncertainty: Heuristics and biases. *Science*, *185*, 1124–1131.

Venn, J. (1880). On the diagrammatic and mechanical representation of propositions and reasonings. *The London, Edinburgh and Dublin Philosophical Magazine and Journal of Science*, *10*(58), 1–18.

Vukotic, V., Pintea, S.-L., Raymond, C., Gravier, G., & Van Gemert, J. C. (2017). One-step time-dependent future video frame prediction with a convolutional encoder-decoder neural network. In *International Conference of Image Analysis and Processing (ICIAP), Proceedings of the 19th International Conference of Image Analysis and Processing*, Catania, Italy.

Wittgenstein, L. (1953). *Philosophical investigations*. Oxford: Basil Blackwell.

Xiong, H. Y., Alipanahi, B., Lee, L. J., Bretschneider, H., Merico, D., Yuen, R. K. C., et al. (2015). The human splicing code reveals new insights into the genetic determinants of disease. *Science,347*(6218).

Zadeh, L. A. (1965). Fuzzy sets. *Informations and Control, 8*, 338–353.

Chapter 2
The Gap Between Symbolic and Connectionist Approaches

2.1 Differences Between Symbolic and Connectionist Approaches

For symbolists, the way of thinking can be fully symbolically simulated without biological embodiment. For connectionists, biological embodiment is a must, and they use connectionist networks for embodiments.

2.1.1 Different Styles of Learning

In symbolic systems, new information is acquired through deduction (or rewriting), starting with some primitives given by human experts. It is a style of *learning by being told* (Dinsmore 1992). The range of acquired information is restricted by the given information (what has been told). An example would be Connection Calculus (Clarke 1981, 1985; Randell et al. 1992), in which regions and their spatial relations are governed only by two axioms as follows.

$$\forall X[\mathbf{C}(X, X)] \tag{2.1}$$

$$\forall XY[\mathbf{C}(X, Y) \rightarrow \mathbf{C}(Y, X)] \tag{2.2}$$

Axiom 2.1 is read as *for any region X, X connects with itself*; Axiom 2.2 is read as *for any region X and Y, if X connects with Y, then Y connects with X*. From the two primitives, we can develop other spatial relations. For example, the relation that region X disconnects from region Y, $\mathbf{DC}(X, Y)$, is defined as that X does not connect with Y, i.e., $\neg\mathbf{C}(X, Y)$. The relation that X is a part of Y, $\mathbf{P}(X, Y)$, is defined as that for any region Z, if Z connects with X, then Z connects with Y, i.e.,

T. Dong, *A Geometric Approach to the Unification of Symbolic Structures and Neural Networks*, Studies in Computational Intelligence 910,
https://doi.org/10.1007/978-3-030-56275-5_2

$\forall Z[\mathbf{C}(Z, X) \rightarrow \mathbf{C}(Z, Y)]$, but neither distance relations, nor orientation relations can be defined without introducing a new axiom as follows.

$$\forall XY[\mathbf{C}(X, Y) \rightarrow \forall \mathfrak{Z} \exists Z[Z \in \mathfrak{Z} \wedge \mathbf{C}(X, Z) \wedge \mathbf{C}(Y, Z)]] \quad (2.3)$$

which is read as *for any region X and Y, if X connects with Y, then for any category* $\mathfrak{Z}$, *there is a member Z in* $\mathfrak{Z}$ *such that Z connects with both X and Y* (Dong 2007, 2008).

To acquire different kind of qualitative orientation relations (e.g., Frank 1991; Freksa 1992; Renz and Mitra 2004), sides of regions and methods of distance comparison must be explicitly introduced (Dong and Guesgen 2007; Dong 2012). For example, let S be a square with vertices A, B, C, and D; l_{AB}, l_{BC}, l_{CD}, and l_{AD} be straight lines that pass AB, BC, CD, and AD, respectively, as illustrated in Fig. 2.1a; let P be a point external to S. By comparing the distance from P to the four sides of S, we can partition the external space of S into 8 areas: $S_A, S_B, S_C, S_D, S_{AB}, S_{BC}, S_{CD}, S_{AD}$, as illustrated in Fig. 2.1b. This method can directly generate other qualitative orientation frameworks. When two opposite sides of square S decrease to 0, we will have the double cross orientation framework (Freksa 1992), as illustrated in Fig. 2.1c; when four sides of square S decrease to 0, we will have a more familiar qualitative orientation framework (Frank 1992), as illustrated in Fig. 2.1d.

In contrast, knowledge in connectionist networks is represented by weights of a network, which can be updated by the back-propagation method (Rumelhart et al. 1986). Back-propagation is a supervised method to update parameters in the way of reducing errors measured by some objective functions. As long as training data are available, parameters can be geared to establish a numeric harmony between the input-output relations in the training data.

2.1.2 Combinatorial Syntax and Semantics or Not?

In symbolic systems, representation and computation are carried out at the same syntactic level; and symbols must have explicit denotation, either something in the external world (extensionalist semantics), or something in the mind (internalist semantics). The denotation of constructed syntactic structure is a function of the denotation of its syntactic parts. Such *combinatorial* feature does not exist for connectionist systems, in which the computation level lies beneath the representation level (Chalmers 1992). Either one node (localist connectionist) or an array of nodes (distributed connectionist) represents a physical or conceptual entity. Semantic information is represented by a pattern in terms of weighted connections. The aim of the forward and backward computations is to update weights in order to reach a harmony in terms of a local minimum (Dinsmore 1992).

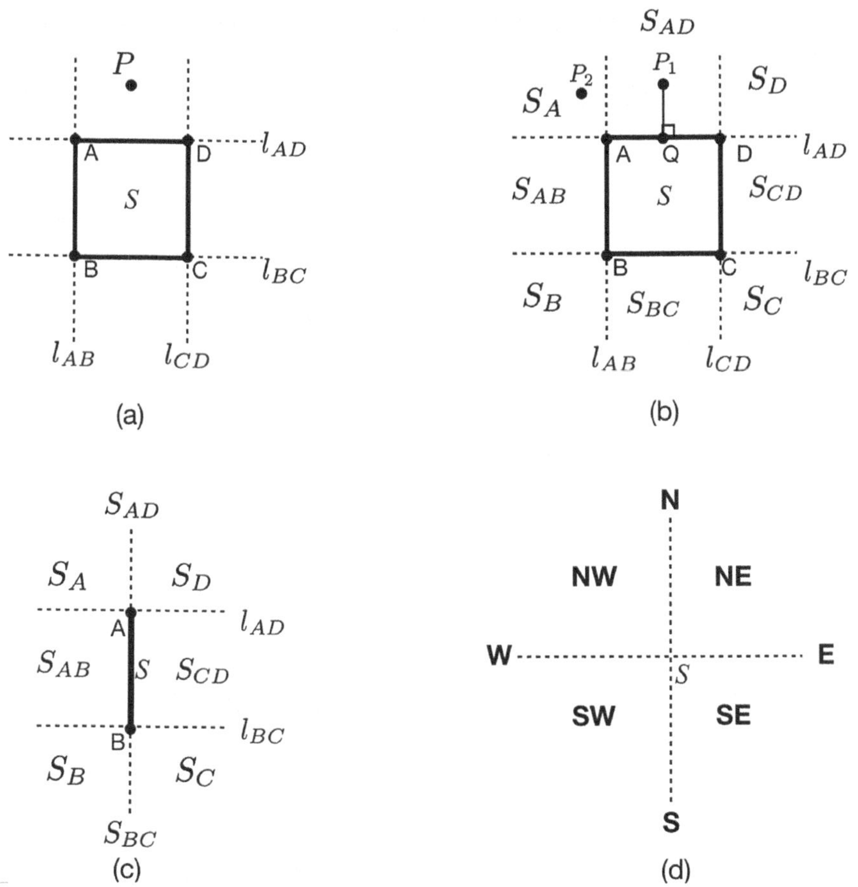

Fig. 2.1 **a** Square S and point P; **b** the distance between P and AD is PQ, the distance between P and AB is PA, so P is nearer to AD than to AB; **c** the double cross orientation framework; **d** a familiar qualitative orientation framework, **N**, **W**, **S**, **E** represent North, West, South, East, respectively

2.1.3 *Structured Processing or Distributed Representation?*

In symbolic systems, the structured representation allows a rule to transfer any expected structure into a structure in a combinatorial way. For example, given a rule that can transfer any structure in the form of $P \wedge Q$ into a new structure in the form of P, we can apply for the structure $[A \vee B \vee C] \wedge [D \vee E \vee F]$, and transform it into $A \vee B \vee C$ (Fodor and Pylyshyn 1988, p. 13). The semantics of $P \wedge Q$ is constructed by its parts, P and Q, along with the operator $\wedge$, and the rule above is about the relation between the whole and its parts—any feature of the whole is also of its parts.

In contrast, in the connectionist model, if a node labeled with $P \wedge Q$ activates a node labeled with P, we only know some causally connected relation between $P \wedge Q$ and P, no part-whole relation can be deducted. That is, connections in networks do not have semantically interpreted components. Connectionist networks are only specifications of causal relations. Node X connecting with Node Y means and only means that the state of node X casually affects the state of node Y (Fodor and Pylyshyn 1988, p. 18). Entities are represented by vectors[1] that ideally satisfy the condition that similar entities have similar vectors. In this way, these vectors serve as distributed micro-feature representations of conceptual level units (Smolensky 1988). These micro-feature vector representations, learned by data, do not have combinatorial structures. The relation between micro-feature and its upper-level concept is just causal effects, or co-occurrence relation, appearing in the dataset. The part-whole relation, even existing in the represented external world, is not encoded by the connectionist network. As *real constituency does have to do with parts and whole* (Fodor and Pylyshyn 1988, p. 22), distributed representations are not a language of thought (Smolensky 1988).

2.2 The Shore of the Gap

Despite the differences, the two approaches do share a number of common features that serve as the shore of the gap.

2.2.1 *Connectionists are* Representationalists, *Too*

Connectionists are *representationalists*, as they hold that there are mental representation for external worlds, which are essential to cognition (Fodor and Pylyshyn 1988). This is also directly evidenced by Rumelhart et al. (1986) such that (PDPs) *are explicitly concerned with the problem of internal representation...strongly committed to the study of representation and process*. Different from the representation of symbolic approaches, the representation of connectionists is at the subsymbolic level. The color world is a good example to understand representations at symbolic and subsymbolic levels. At the symbolic level, we have terms such as "red", "green", "yellow", "pink"; while at the subsymbolic level, we have continuous physical features, such as wavelength, intensity, or reflectance. However, mapping between the two worlds is not trivial (Rosch 1973, p. 329).

[1] In *localist* representation, an entity (a concept) is represented by one node, which can be viewed as a one-element vector.

2.2.2 *No Significant Difference in Computational Architectures*

There is no fundamental differences in computational architectures underlying symbolic and connectionist computing models (Adams et al. 1992), as Turing machines can be implemented either by symbolic models or by connectionist networks. Especially, localist networks, in which a concept is represented by one node, can be simply regarded as special symbolic models without rules.

2.2.3 *Connectionists are Functionalists*

Though inspired by the architecture of brain, connectionist models focus on functional simulation of the mind. Connectionist networks often have mechanisms that do not have neural counterparts and that structures and processes found in the brain do not have simulating counterparts in the connectionist network models (Aizawa 1992). For example, neither back propagation nor error correction operations exist in the olfactory system or elsewhere in biological brains (Freeman 1988).

2.3 Philosophies for the Gap

With the well-known thought experiment of Chinese Room, Searle (1980) refuted that symbolic systems alone can capture vital features of the mind. Pinker and Prince (1988) listed three possible philosophies for the relation between connectionist and symbolic approaches, namely, eliminative connectionism (*eliminativism*), implementational connectionism (*implementationalism*), and revisionist connectionism (*revisionism*).

Observing that symbolic approach is inadequate to account human cognition and that connectionist networks are inspired by the physiology of the mind, people might believe that connectionist approaches can achieve all symbolic approaches can do, and coin the term *eliminativism*, in the sense that connectionist approaches can sweep away (*eliminate*) symbolic approaches. Chalmers (1992) refuted *eliminativism* by arguing that there is no profound computational differences between connectionist networks and symbolic computational devices, as they can be simulated by each other. By dropping *eliminativism*, people see the gap between connectionist and symbolic approaches, and have been attempting to close it for decades.

One possibility is *implementationalism* that holds the view that connectionist networks are an alternative way to implement symbolic computing. That is, symbolic system is a virtual machine, and connectionist network is the hardware of this virtual machine. Smolensky (1988) refuted *implementationalism* for the reason that cognition cannot be adequately described at the symbolic level; Dyer (1988) argued that if symbolic operations were the target of connectionist models, a variety of useful cognitive properties of connectionist models will be lost at symbolic level. A weaker version is *revisionism* that holds that a symbolic account can be generated by connectionist networks and that connectionist networks can also lead to different symbolic structures and operations. However, *revisionism* does not point out different ways of *revision*.

Inspired by the relation between statistical thermodynamics and symbolic descriptions about temperature and heat, Smolensky (1988) coined the term *limitivism* that accurate connectionist account can approximate good symbolic descriptions within certain *limit*. Then, the gap between symbolic and connectionist approaches can only be narrowed, and cannot be closed. So, a *patchwork*, or *stepstone*, for the gap is needed, that is the *hybridism* which is favored by a number of researchers. Dinsmore (1992) suggested that successful cognitive systems shall accept both connectionist and symbolic perspectives, and should be capable of solving problems from both perspectives.

2.4 Connectionists' Attempts to Represent Symbolic Structures

In symbolic approaches, concepts are either primitive, or constructed using primitives. Early connectionists' attempts to represent these symbolic concepts were either to represent each concept as a single unit (Barlow 1972), or latent features as a fixed-length of array (Willshaw et al. 1969; Hopfield 1982). The advantage of feature vector representation is good for reasoning with similarity relations, however, not for structural relations. *The real challenge for connectionists will not be to defeat symbolic theorists, but rather to come to terms with the ongoing relevance of the symbolic level of analysis* (Bechtel and Abrahamsen 2002, p. 16).

2.4.1 Representing Propositions

Representing propositions using networks can date back to Hinton (1981) who used four groups (called *assemblies*) to approximate a Triple relation[2] as follows: the first three groups represent the three components of a Triple, the forth group, namely PROP, represents the Triple as a whole, under the condition that similar Triple components have similar PROP representation. This is achieved by repeatedly updating the units of the four groups. Two side-effects are as follows: (1) a concept occurring in different Triples will have different numeric representations; (2) manual efforts are needed to assign the right entity to the right group. Hinton (1986) improved the model by introducing role-specific units as hidden layers between concept units and

[2]A Triple in knowledge-graph takes the form that consists of (*head, relation, tail*).

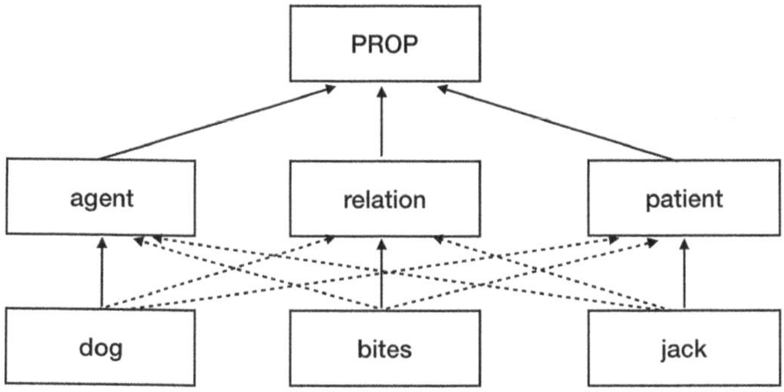

Fig. 2.2 By introducing role-specific units (agent, relation, patient) as the hidden layer between the concept layer and the proposition layer, the network can learn to assign concepts to role units

PROP, so that networks could have the freedom to choose groups for concepts, as illustrated in Fig. 2.2. Compared with pure symbolic Triple representation, network representation encodes both direct content (or micro-features) and associative content (or relations to other concepts) of a concept. This feature is valued in human similarity judgments (Tversky 1977), as well as in the judgment under uncertainty (Tversky and Kahneman 1974).

2.4.2 *Representing Part-Whole Relations*

A part-whole relation of a Triple consists of a part, a whole, and the part-whole relation, e.g., (*downtown, part-of, city*), (*toe, part-of, the body*), (*Roman Empire, part-of, the Italian history*), (*the boundary, part-of, a closed region*), (*your dream, part-of, your mental activities*). To recognize part-whole relations between letters and words, Hinton (1990) proposed *within-level time-sharing connections*, which means that the same letter at different locations has different embedding vectors, as illustrated in Fig. 2.3. Whether letter p is part of apple will be inferred by p's two connections: one between p, the hidden layer and the other between p's position and the hidden layer. When both connections are activated, p will be inferred as part of apple. Hinton (1990) named it *simple intuitive inference* in the sense that inference is performed by a single connectionist network.

The structure of the sentence *Tom knows Ted likes cat* is a hierarchical structure that shares the grammatical knowledge at different layers. Linguistic structure can be nested in arbitrary number of layers, such as *once upon a time, an old monk tells a baby monk a story that once upon a time, an old monk tells a baby monk a story that* ... Hinton (1990) suggested that different structure layers can share the same connectionist network, and proposed *between-level time-sharing connections*, which means that part-whole structures at different levels in a hierarchical tree share the same network structure at different times, as illustrated in Fig. 2.4. The inference is

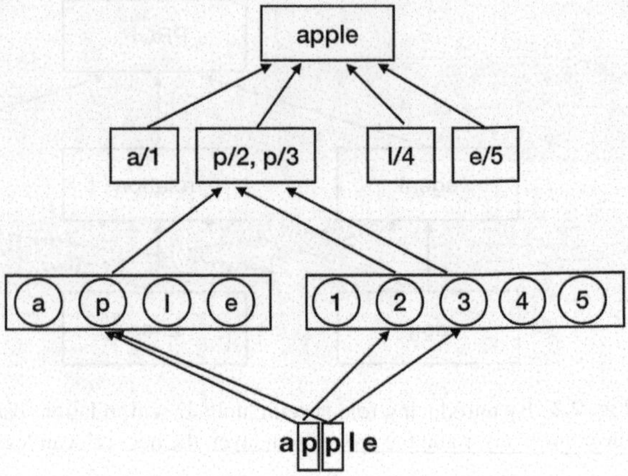

Fig. 2.3 The letter p is part of the word apple. After training process, p will have different vector embeddings, when located at different positions

then performed in a sequence of *simple intuitive inferences*: whether cat is part of the whole sentence can be approximately determined by a sequence of two *simple intuitive inferences*: whether cat is part of S2 and whether S2 is part of S1. Such inference is called *rational inference*.

2.4.3 Representing Tree Structures

The data structures in symbolic approaches are versatile, such as lists, trees, graphs. Closing the gap between connectionist and symbolic approaches raises the question on how to represent symbolic structures by connectionist networks? Following the distributed convention, Pollack (1990) developed RAAM (Recursive Auto-Associative Memory) model to encode symbolic tree structures. RAAM initializes leaf nodes with numeric vectors, and uses three-layered auto-associative encoder networks to recursively train other nodes, as illustrated in Fig. 2.5. RAAM is able to create numeric vectors for all non-leaf nodes following the criteria that similar structures have similar numeric vectors. Pollack (1990) showed that RAAM can cluster a variety of tree structures, such as words of different lengths, syntactic trees, sentences in the form of ternary trees.

RAAM has a number of weakness, for example, it needs lots of epochs for training, the size of the hidden layer can be large, the vectorial representation is biased by more recent training data. Adamson and Damper (1999) proposed B-RAAM to solve these problems. To make the hidden information persist in a longer time period and to increase its effect on training, Adamson and Damper (1999) introduced a delay line as part of the input layer to store previous hidden layer, and concatenated past input symbols at the output layer. A sub-network is trained together with the hidden layer and activated nodes in the output layer, so that outputs can be correctly interpreted, as illustrated in Fig. 2.6. A series of experiments showed that B-RAAM demonstrates

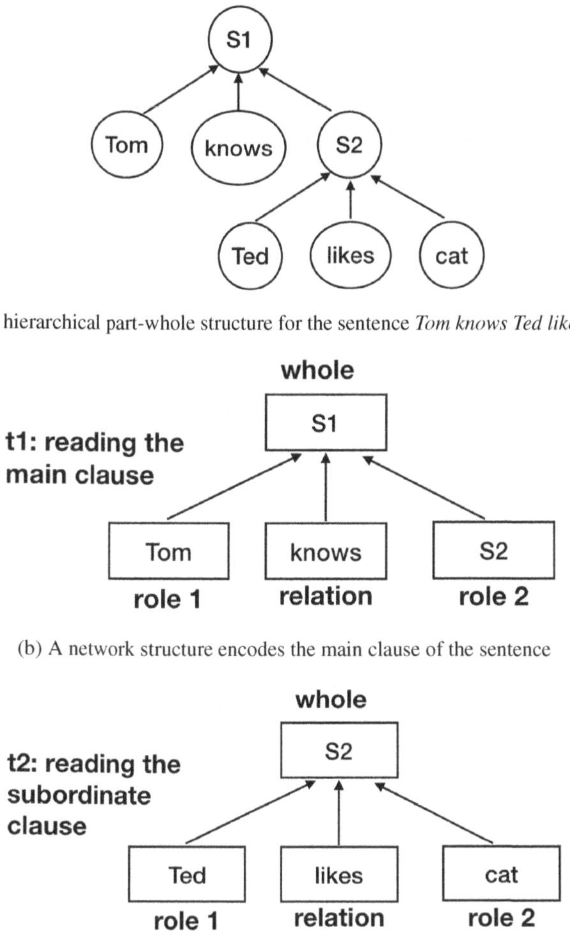

(a) A hierarchical part-whole structure for the sentence *Tom knows Ted likes cat*

(b) A network structure encodes the main clause of the sentence

(c) The same network structure encodes the subordinate clause of the sentence

Fig. 2.4 Mapping different part-whole relations into the same network structure, S2 has different embedding vectors in (**b**) and (**c**)

better performance than RAAM, such as B-RAAM has fewer training epochs, smaller size of the hidden layer.

Representation and searching are two appealing capabilities of symbolic approaches, which challenge connectionists to apply connectionist networks for knowledge representation and searching tasks. Giles and Gori (1998) learned finite-state automaton using back-propagation. However, it is not easy for connectionist networks to represent sophisticated knowledge representation. Sun and Bookman (1994) surveyed a number of connectionist methods that approximated logical inferences and searching. Almost all the connectionist approaches in the literature follow

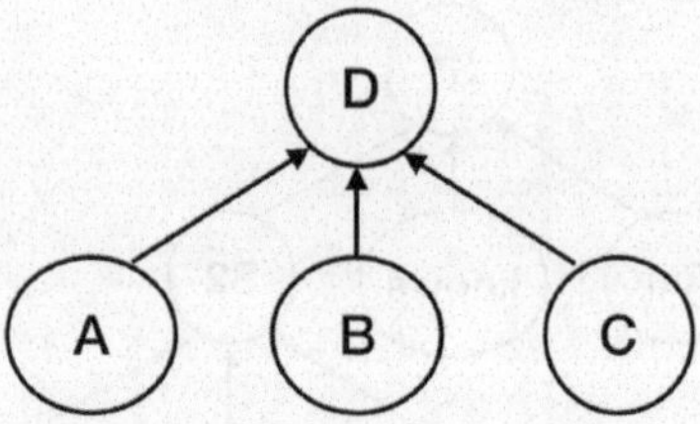

(a) A tree whose leaf nodes A, B, C have numeric vector representations

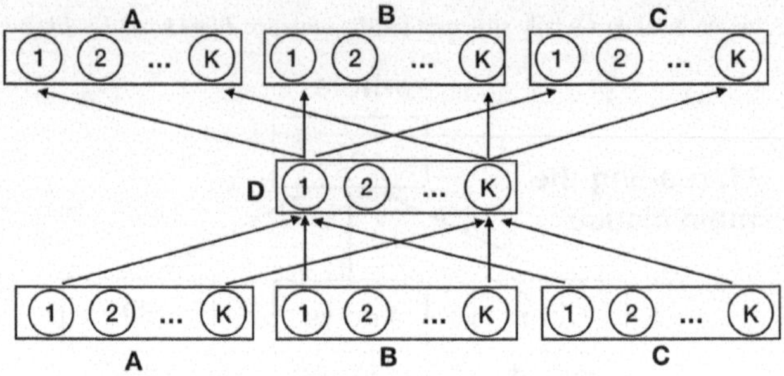

(b) A three-layered auto-associative encoder network

Fig. 2.5 **a** A tree structure in which A, B, C have vector embeddings, D does not; **b** we can compute the vector embedding of D by using back-propagation algorithm

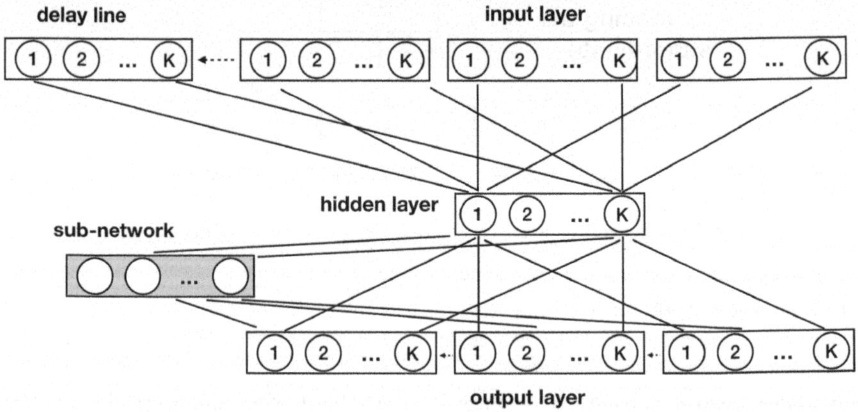

Fig. 2.6 The architecture of B-RAAM

early ideas advocated by Willshaw et al. (1969), Hopfield (1982), Hinton (1986). They represent a concept either as a single unit with no internal structure or as a feature vector, use connections to encode relations, and train these connections by utilizing the back-propagation methods (Rumelhart et al. 1986). This can be evidenced by a variety of international conferences or workshops, such as Bader et al.

(2007), Besold et al. (2017), as well as the NeSy[3] series from 2005 to 2019. They fall within the philosophy of *limitivism*, attempting to design better connectionist architectures, and to approximate symbolic reasoning and decision making within a certain limit.

2.5 Hybrid Systems

The complementary competence of connectionist and symbolic approaches have intrigued researchers to develop hybrid models (Sun and Bookman 1994; Sun and Alexandre 1997; Sun 2002, 2016), or even to develop unified models (Newell 1990). It turns out that the architecture of a hybrid model is more complicated than expected, and shall be carefully and very artfully designed, because such models consist of a number of different types of representations and procedures. Learning of hybrid models is more difficult than learning of connectionist models, because hybrid models must perform the learning of symbolic components that are notoriously hard. In an attempt to synergistically integrate symbolic and connectionist components, Sun and Peterson (1998) show that symbolic knowledge can be learned from subsymbolic knowledge and argue for developing new learning methods to achieve an intimate and synergistic combination of symbolic and subsymbolic learning process.

2.6 Summary

In this chapter, we reviewed connectionist and symbolic approaches, surveyed a number of philosophies for closing the gap between the two approaches, revisited connectionists' efforts to represent symbolic structures. We end with an observation that new methods are expected to develop better hybrid models.

References

Adams, F., Aizawa, K., & Fuller, G. (1992). Rules in programming languages and networks. In *The symbolic and connectionist paradigms: Closing the gap* (pp. 49–68). Hillsdale NJ: Erlbaum.

Adamson, M. J., & Damper, R. I. (1999). B-RAAM: A connectionist model which develops holistic internal representations of symbolic structures. *Connection Science*, *11*(1), 41–71.

Aizawa, K. (1992). Biology and sufficiency in connectionist theory. In *The symbolic and connectionist paradigms: Closing the gap* (pp. 69–88). Hillsdale, NJ: Erlbaum.

Bader, S., Hitzler, P., Hölldobler, S., & Witzel, A. (2007). A fully connectionist model generator for covered first-order logic programs. In *IJCAI 2007, Proceedings of the 20th International Joint Conference on Artificial Intelligence* (pp. 666–671), Hyderabad, India, 6–12 January 2007.

[3]Neural-Symbolic Learning and Reasoning http://www.neural-symbolic.org/.

Barlow, H. (1972). Single units and sensation: A neuron doctrine for perceptual psychology? *Perception, 1*, 371–394.
Bechtel, W., & Abrahamsen, A. (2002). *Connectionism and the mind: Parallel processing, dynamics, and evolution in networks*. Hong Kong: Graphicraft Ltd.
Besold, T. R., d'Avila Garcez, A., & Lamb, L. C. (2017). Human-like neural-symbolic computing (Dagstuhl Seminar 17192). *Dagstuhl Reports*, *7*(5), 56–83.
Chalmers, D. J. (1992). Subsymbolic computation and the Chinese room. In *The symbolic and connectionist paradigms: Closing the gap* (pp. 25–48). Hillsdale, NJ: Erlbaum.
Clarke, B. L. (1981). A calculus of individuals based on 'connection'. *Notre Dame Journal of Formal Logic*, *23*(3), 204–218.
Clarke, B. L. (1985). Individuals and points. *Notre Dame Journal of Formal Logic*, *26*(1), 61–75.
Dinsmore, J. (1992). Thunder in the gap. In *The symbolic and connectionist paradigms: Closing the gap* (pp. 1–23). Erlbaum.
Dong, T. (2007). The nine comments on the RCC theory. *AAAI'07 Workshop on Spatial and Temporal Reasoning* (pp. 16–20). Canada: Vancouver.
Dong, T. (2008). A comment on RCC: From RCC to RCC^{++}. *Journal of Philosophical Logic*, *37*(4), 319–352.
Dong, T. (2012). *Recognizing variable environments—The theory of cognitive prism*, volume 388 of Studies in computational intelligence. Berlin, Heidelberg: Springer.
Dong, T., & Guesgen, H. (2007). Is an orientation relation a distance comparison relation? In *IJCAI'07 Workshop on Spatial and Temporal Reasoning* (pp. 45–51), Hyderabad, India.
Dyer, M. G. (1988). The promise and problems of connectionism. *Behavioral and Brain Sciences*, *1*, 32–33.
Fodor, J. A., & Pylyshyn, Z. W. (1988). Connectionism and cognitive architecture—A critical analysis. *Cognition*, *28*(1–2), 3–71.
Frank, A. (1991). Qualitative spatial reasoning with cardinal directions. In *Proceedings of the Seventh Austrian Conference on Artificial Intelligence* (pp. 157–167). Berlin, Wien: Springer.
Frank, A. (1992). Qualitative spatial reasoning about distances and orientations in geographic space. *Journal of Visual Language and Computing*, *3*, 343–371.
Freeman, W. J. (1988). Dynamic systems and the "subsymbolic level". *Behavioral and Brain Sciences*, *1*, 33–34.
Freksa, C. (1992). Using orientation information for qualitative spatial reasoning. In *Proceedings of the International Conference GIS-From Space to Territory: Theories and Methods of Spatial-Temporal Reasoning*, LNCS. Pisa: Springer.
Giles, C. L. and Gori, M. (Eds.). (1998). *Adaptive processing of sequences and data structures, international summer school on neural networks, "E.R. Caianiello"—Tutorial lectures*, London, UK, UK: Springer.
Hinton, G. E. (1981). Implementing semantic networks in parallel hardware. In G. E. Hinton & J. A. Anderson (Eds.), *Parallel models of associative memory* (pp. 161–187). Hillsdale, NJ: Erlbaum.
Hinton, G. E. (1986). Learning distributed representations of concepts. In *Proceedings of the Eighth Annual Conference of the Cognitive Science Society* (Vol. 1, p. 12), Amherst, MA.
Hinton, G. E. (1990). Mapping part-whole hierarchies into connectionist networks. *Artificial Intelligence*, *46*(1–2), 47–75.
Hopfield, J. J. (1982). Neural networks and physical systems with emergent collective computational abilities. *Proceedings of the National Academy of Sciences of the United States of America*, *79*(8), 2554–2558.
Newell, A. (1990). *Unified theories of cognition*. Cambridge, MA: Harvard University Press.
Pinker, S., & Prince, A. (1988). On language and connectionism: Analysis of a parallel distributed processing model of language acquisition. *Cognition, 28*, 73–193.
Pollack, J. B. (1990). Recursive distributed representations. *Artificial Intelligence*, *46*(1–2), 77–105.
Randell, D., Cui, Z., & Cohn, A. (1992). A spatial logic based on regions and connection. In B. Nebel, W. Swartout, & C. Rich (Eds.), *Proceeding of the 3rd International Conference on Knowledge Representation and Reasoning* (pp. 165–176). San Mateo: Morgan Kaufmann.

Renz, J., & Mitra, D. (2004). Qualitative direction calculi with arbitrary granularity. In C. Zhang, H. Guesgen, & W. Yeap (Eds.), *PRICAI 2004: Trends in Artificial Intelligence: 8th Pacific Rim International Conference on Artificial Intelligence* (pp. 65–74). Berlin, Heidelberg and New Zealand: Springer.

Rosch, E. (1973). Natural categories. *Cognitive Psychology*, *3*, 328–350.

Rumelhart, D. E., McClelland, J. L., & PDP Research Group (Eds.). (1986). *Parallel distributed processing: Explorations in the microstructure of cognition, Vol. 1: Foundations*. Cambridge, MA, USA: MIT Press.

Searle, J. (1980). Minds, brains, and programs. *Behavioral and Brain Sciences*, *3*(3), 417–457.

Smolensky, P. (1988). On the proper treatment of connectionism. *Behavioral and Brain Sciences*, *1*, 1–23.

Sun, R. (2002). Hybrid connectionist symbolic systems. In M. Arbib (Ed.), *Handbook of brain theories and neural networks* (2nd ed., pp. 543–547). Cambridge, MA: MIT Press.

Sun, R. (2016). Implicit and explicit processes: Their relation, interaction, and competition. In L. Macchi, M. Bagassi, & R. Viale (Eds.), *Cognitive unconscious and human rationality* (pp. 27–52). Cambridge, MA: MIT Press.

Sun, R., & Alexandre, F. (Eds.). (1997). *Connectionist-symbolic integration: From unified to hybrid approaches*. Hillsdale, NJ, USA: L. Erlbaum Associates Inc.

Sun, R., & Bookman, L. A. (1994). *Computational architectures integrating neural and symbolic processes: A perspective on the state of the art*. Norwell, MA, USA: Kluwer Academic Publishers.

Sun, R., & Peterson, T. (1998). Autonomous learning of sequential tasks: Experiments and analyses. *IEEE Transactions on Neural Networks*, *9*(6), 1217–1234.

Tversky, A. (1977). Features of similarity. *Psychological Review*, *84*, 327–353.

Tversky, A., & Kahneman, D. (1974). Judgment under uncertainty: Heuristics and biases. *Science*, *185*, 1124–1131.

Willshaw, D. J., Buneman, O. P., & Longuet-Higgins, H. C. (1969). Non-holographic associative memory. *Nature*, *222*, 960–962.

Chapter 3
Spatializing Symbolic Structures for the Gap

> *I advocate a third form of representing information that is based on using geometrical structures rather than symbols or connections among neurons.*
> —(Gärdenfors 2000, p. 2)

3.1 System 1 of the Mind

The function of the mind consists of two different types, namely, System 1 and System 2 (Kahneman 2011). The basic function of System 1 (fast thinking) is *associative activation* (Kahneman 2011, pp. 50–52), with which evoked ideas continue to evoke many other coherent ideas, in a spreading cascade way. System 1 is powerful to automatically associate almost everything to construct a harmony story. Look at the two words in Fig. 3.1, you will experience some unpleasant memories, and your face will be twisted a little bit, and you may even pull back your head. Your heart beats a bit faster (Kahneman 2011, p. 50). The secret behind your reactions is that System 1 of your mind automatically constructs a temporal sequence and a causal connection between **banana** and **vomit**, forming a scenario that **banana** caused **sickness**, which further caused **vomit**. System 1 has the power of evoking coherent ideas in a cascade way: a word evokes memories, memories triggers emotions,[1] emotions in turn evoke facial expressions, facial expressions intensify the feelings to which they link …

Now, try to fill a letter to complete the word fragment **SO_P** (Kahneman 2011, p. 52), as shown in Fig. 3.2. It is very likely that you will fill **U** to make the word **SOUP**,

[1]The Third Law of Cognition: Feeling comes first (Tversky 2019, p. 42).

T. Dong, *A Geometric Approach to the Unification of Symbolic Structures and Neural Networks*, Studies in Computational Intelligence 910,
https://doi.org/10.1007/978-3-030-56275-5_3

Bananas Vomit

Fig. 3.1 If you look at the two words **Bananas** and **vomit**, you may not feel very comfortable (Kahneman 2011, p. 50)

SO_P

Fig. 3.2 Filling a letter between **O** and **P** to form a word (Kahneman 2011, p. 52)

instead of filling **A** for **SOAP**. Because your mind is unconsciously affected by the **bananas** and **vomit** example, forming a bit more detailed scenario that drinking some **soup** after eating **banana** caused sickness. Current task is primed by the earlier one. This is the *priming effect* triggered by System 1, which goes deeper to the subconcept level, as illustrated by the following experiments (Kahneman 2011, pp. 52–58).

Two groups of students play word assembly game: given 5 words, they need to order them into a meaningful sentence. For example, words for Group One can be *finds, he, it, yellow, instantly*, words for Group Two can be *Florida, forgetful, in, she, became*. After finishing this task, two groups are asked to walk to another office, to do another experiment. The aim of this experiment is this walk—Student in Group Two walked significantly more slowly than others. The reason is that Group Two played with words associated with the elderly, such as *Florida, forgetful, bald, grey, wrinkle*. These words prime thoughts of old age, though the word *old* is never mentioned, and further prime the associated behavior—walking slowly. This is the *Florida effect* of System 1 (Kahneman 2011, p. 53).

System 1 is even able to generate simpler and more coherent things than the real. If you like a girl's appearance, you will probably like everything of her; if you dislike the appearance of a beggar, you will probably dislike his everything. If you like the president's politics, you will probably like his voice and his appearance as well. System 1 has the tendency to like (or dislike) everything about a person—including things you have not observed. This special *priming effect* is the *halo effect* that exaggerates emotional coherence. *Halo effect* plays a large role in shaping our view and decision making of people and situations (Kahneman 2011, p. 82).

System 1 assumes that current available information is all that needs to know, and uses them as a base to construct a coherent model. Its slogan is "What You See Is All There IS" (WYSIATI) (Kahneman 2011, p. 85). The success for System 1 is measured by the coherence of the story it manages to create, irrelevant to the amount and the quality of the data it has. WYSIATI facilitates the achievement of both coherence and the cognitive ease that causes us to accept a statement as true. It explains why we can think fast, and how we are able to make sense of partial information in a complex world.[2] With WYSIATI, System 1 is capable of constructing the best possible story that incorporates ideas current activated, and deal with stories in which the elements are causally linked, but System 1 is weak in reasoning (Fig. 3.3).

[2] The Seventh Law of Cognition: The mind fills in missing information (Tversky 2019, pp. 78, 244).

```
      1 7
    x 2 4
    -----
      2 8
      4
    1 4
  + 2
  -------
    1 0 8
  + 3
  -------
    4 0 8
```

Fig. 3.3 System 2 follows rules to perform calculation

3.2 System 2 of the Mind

System 2 is the mathematical section of the mind, which can follow rules, perform calculation, reasoning, and make deliberate choices. To compute 17×24, you first retrieve from the memory the program for multiplication, then you implement 17×24 (Kahneman 2011, p. 23). You feel the burden of holding much material in memory, need to keep track of where you are, and of where you are going, and need to hold on to the intermediate result. This mental process is deliberate, effortful, and orderly (Kahneman 2011, p. 20). Typical tasks of System 2 listed in (Kahneman 2011, p. 22) are as follow.

- brace for the starter gun in a race
- focus attention on the clowns in the circus, or on the voice of a particular person in a crowded and noisy room
- look for a woman with white hair
- park in a narrow space
- tell someone your telephone number
- maintain a faster walking speed than is natural for you
- check the validity of a complex logical argument.

3.3 Interaction Between System 1 and System 2

With *priming effects*, System 1 believes almost everything automatically and quickly, and can be easily fooled. System 2 is cautious, operates effortful mental activities, including complex computation and logical reasoning. Despite all these merits, System 2 is lazy (Kahneman 2011, p. 48). Consequently, System 2 often works on ideas generated by System 1, and has the ability to change the way System 1 works.

Having a quick glance at images in Fig. 3.4a–c, you may have one or two different interpretations generated as the automatic response of System 1 of your mind. For example, the object in Fig. 3.4a can be either a saxophone player, or the face of a lady; the object in Fig. 3.4b can be either a young girl, or an old woman; the object

(a) a saxophone player or a lady's face?

(b) an old woman or a young girl?

(c) two faces or a vase?

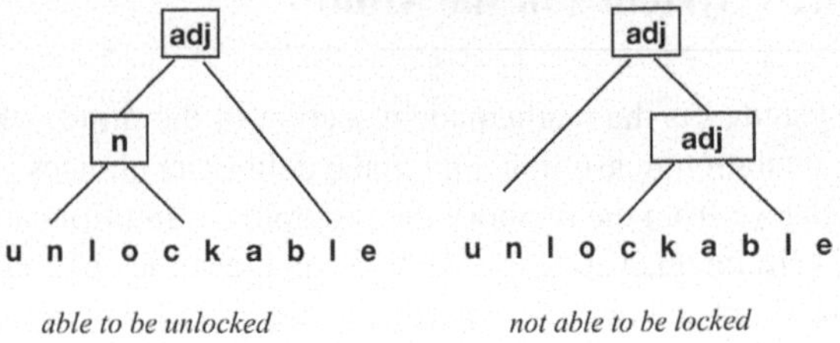

(d) lockable or not?

Fig. 3.4 Quick glancing at these images may have different recognition results

in Fig. 3.4c can be either a vase, or two faces. If I ask you why I fail to recognize Fig. 3.4a as the lady, following your instructions, I will start System 2 of my mind, concentrate myself to search where are the two eyes, the nose and the mouth of the lady. Especially, if you are asked why the word unlockable has two meanings in Fig. 3.4d, you will examine the two tree structures with some effort and concentration.

3.4 Representation and Reasoning

> Physicians, engineers, mechanics and others use errors as signs of malfunctioning, that some system has broken down and is in need of repair. Not so for psychologists. Errors are viewed as natural products of the systems, and as such are clues to the way the system operates.
>
> —(Tversky 1992, p. 131)

Psychologists represent *associative activation* of System 1 with nodes in a vast network (associative memory), in which each node is linked to many others. Quillian (1968) coined the term *Semantic Networks* to represent associative memory. Later, semantic networks turns out to be one of the interesting topic for symbolic and sub-symbolic debate, perhaps because they are *the most connectionist symbolic models* (Dinsmore 1992, p. 3).

Fig. 3.5 San Diego (CA) is further east than Reno (NV). Picture copied from Google Map

Byrne (1979) proposed that spatial relations among objects were mentally represented as a propositional network. For example, a semantic network of an urban environment could be a labeled network of spatial objects. Figure 3.6 illustrates a semantic network that California is located in the west of Nevada and that San Diego is located in California and that Reno is located in Nevada, as shown in Fig. 3.5. Most people mistakingly judge that San Diego is further west than Reno (Stevens and Coupe 1978). To account for these errors, Stevens and Coupe (1978) proposed a partially hierarchical structure of the spatial representation in mind:

- Nodes in the tree are represented by regions;
- Relations between two locations are explicitly stored, if the two locations are in the same spatial region;
- Relations can be inferred by combining spatial relations.

Accordingly, all these spatial objects shall be represented in regions. California and Nevada are located inside the same region USA, and the relation between them is explicitly represented. San Diego is geographically located inside California, Reno is located inside Nevada, the relation between San Diego and Reno is not explicitly represented, as illustrated in Fig. 3.7. With this structure, the WYSIATI effect of System 1 will mistakenly predict that San Diego is also to the west of Reno.

Generally, people make systematically errors in reasoning with spatial knowledge (Tversky 1981; McNamara 1991). For example, people may mistakenly assume that

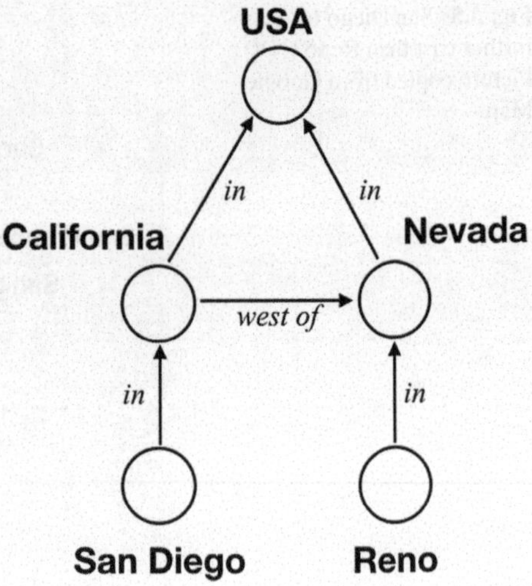

Fig. 3.6 A simple semantic network of spatial knowledge among California, Nevada, San Diego, and Reno

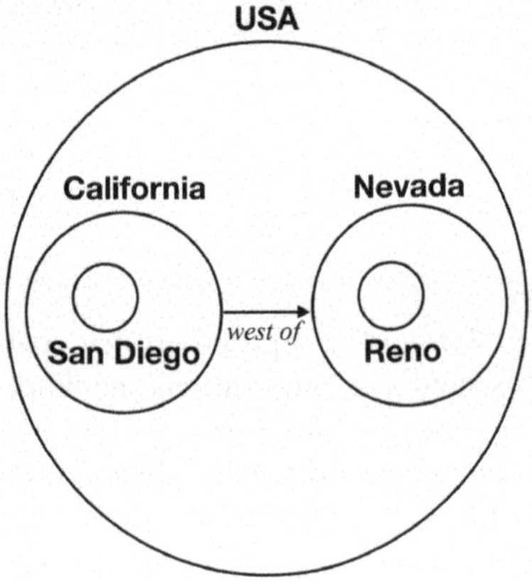

Fig. 3.7 A partially region-based hierarchical structure among California, Nevada, San Diego, and Reno

Madrid (Spain) is further south than Washington (DC) or that Seattle (USA) is further south than Montreal (Canada) or that Nanjing (located in the south of the Yangtze River) is further north than Nan-tong (located in the north of the Yangtze River).

Let us revisit the debate in Sect. 1.2. Suppose that we have developed a network, without imposing structures, to identify the *is-a* relation between two entities. After a long time of training, the network works perfectly well.[3] Given entity representations of *dog* and *animal*, the network will signal *true*, given entity representations of *dog* and *cat*, the network will signal *false*. The question is: if this network works perfectly well, how shall entities and relations be represented? In the literature of representational learning, both entities and relations are represented by vectors. The relation that $(dog, isa, animal)$ is approximated by the relation $\mathbf{dog} + \mathbf{isa} \approx \mathbf{animal}$, which can be understood as a *translation* from **dog** to **animal** using **isa** (Bordes

[3]Dyer (1988) called this "connectoplasm".

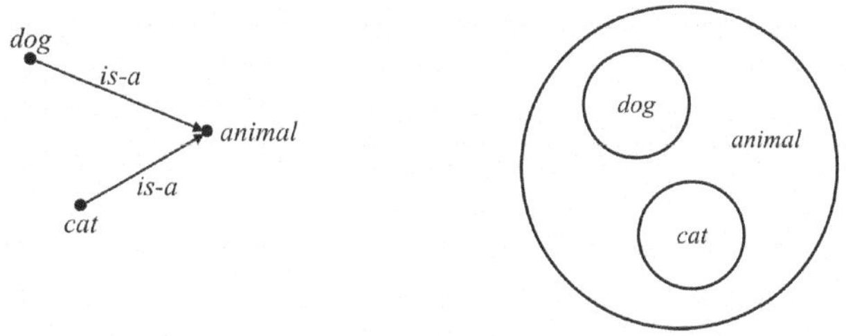

(a) Entities and relations are explicitly represented by vectors

(b) Entities are represented by regions, the *is-a* relationis not explicitly represented

Fig. 3.8 How shall we perfectly represent the *is-a* relation?

et al. 2013). The limitation of the translation-based approach is that all relations have to be explicitly represented (Bordes et al. 2013; Wang et al. 2014; Lin et al. 2015; Ji et al. 2015). It is not difficult to understand that perfect vector representation for the *is-a* relation does not exist: Let **isa** be the perfect vector representation for the *is-a* relation, $(dog, isa, animal)$ and $(cat, isa, animal)$ be two assertions in the Triple form, as illustrated in Fig. 3.8a. Ideally, we have $\mathbf{dog} + \mathbf{isa} = \mathbf{animal}$ and $\mathbf{cat} + \mathbf{isa} = \mathbf{animal}$, and we will have $\mathbf{dog} = \mathbf{cat} = \mathbf{animal} - \mathbf{isa}$. The interpretation of $\mathbf{dog} = \mathbf{cat}$ is that dog and cat are the same entity. We propose not to explicitly represent the *is-a* relation as a vector, rather that to represent entities as regions, so that the *is-a* relation can be implicitly and precisely encoded by region configuration that *dog* region and *cat* region are located inside *animal* region, as illustrated in Fig. 3.8b.

3.5 Precisely Spatializing Symbolic Structures onto Vector Space

> It has been a common prejudice in cognitive science that the brain is either a Turing machine working with symbols or a connectionist system using neural-networks.
>
> —(Gärdenfors 2000, p. 2)

If we use connectionist approaches to simulating the associative function of System 1,[4] and symbolic approaches to simulate System 2, we will have to figure out a way that System 2 can affect System 1 (see Sect. 3.3). Normally, a symbolic system is reviewed as something of *disembodied abstractness* (Dreyfus et al. 1986; Sun 2015). The precondition for a symbolic system to affect connectionist approach is that symbolic structures can be precisely embodied in the vector space. Our method is to promote vector outputs from connectionist networks into regions under the condition

[4]Connectionism is a special case of associationism that models associations using artificial neuron networks (Gärdenfors 2000, p. 1).

that spatial relations among these regions precisely encode symbolic structures. This looks like the well-known *symbolic grounding problem* (Harnad 1990, 2003), which addresses the problem on finding an intrinsic interpretation of a symbol system within the target environment. This problem is important in robotics, and receives continued interests in tasks related with human-robot interactions (Steels 2008; Hristov et al. 2017). In our case, we can state the problem as follows: how can we make the semantic interpretation of a symbolic structure intrinsic to the vector space, while vectors from connectionist networks are well-preserved in this semantic interpretation? A series of experiments, by other researchers (Erk 2009; Fu et al. 2015; Faruqui et al. 2015; Li et al. 2019) and by ourselves (Dong et al. 2019b), show that it is impossible to precisely ground large symbolic tree structures into vectors within the space structured by connectionist networks. However, it is possible to spatialize symbols as regions in higher dimensional space (Dong et al. 2019a, b). Therefore, we generalize the symbol grounding problem to the *symbol spatialization problem*. Formally, we describe it as follows.

Definition Let symbolic structure $\mathfrak{S}$ be a relational structure (T, S) in which T is a set of symbols $\{t_0, t_1, \ldots\}$ and S is a set of relations $\{s_0, s_1, \ldots\}$. Let $\mathfrak{V} : T \rightarrow \mathbb{R}^m$ be a vector embedding function, $\mathfrak{W} : T \rightarrow \mathbb{R}^n$ be a vector extension function, Ξ be a set of spatial relations $\{\xi_{s_i}\}$ with the condition that $s_i \in S$ is grounded on (or diagrammed by) ξ_{s_i}, written as $\psi : S \rightarrow \Xi$.

A promoting operator $\Theta_0 : T \times \mathfrak{V} \times \mathfrak{W} \rightarrow \bigodot$ initializes $t_i \in T$ with an $(m+n)$-dimensional ball $\odot(O_{t_i}, r_{t_i}) \in \bigodot$, O_{t_i} is the center point of the ball, radius $r_{t_i} = 0^+$, $O_{t_i}[0:m] = \mathfrak{V}(t_i)$, and $O_{t_i}[m:m+n] = \mathfrak{W}(t_i)$. $\odot(O_{t_i}, r_{t_i})$ is called $\mathcal{N}$-Ball.

A grounding (or diagramming) operator $\Theta_1 : \bigodot \times S \times \psi \rightarrow \bigodot$ gears $O_{t_i}[m : m+n]$ and radius r_{t_i} so that s_i is precisely grounded on (or diagrammed by) $\psi(s_i) = \xi_{s_i}$.

A spatializing operator Θ spatializes $\mathfrak{S}$ onto an $\mathcal{N}$-Ball in $(m+n)$-dimensional space, while well-preserving $\mathfrak{V}$.

$$\Theta(\mathfrak{S}, \mathfrak{V}, \mathfrak{W}, \psi) = \Theta_1(\Theta_0(T, \mathfrak{V}, \mathfrak{W}), S, \psi)$$

in which, Θ_0 is the promoting operator that promotes a vector into an $\mathcal{N}$-Ball in higher dimension space; its inverse Θ_0^{-1} is the projecting operator. Θ_1 is the grounding or diagramming operator that grounds a symbolic structure onto a configuration of $\mathcal{N}$-Balls; its inverse is the abstracting operator Θ_1^{-1}.

Here, a symbol is spatialized into a ball in a high dimensional space with the restriction that the center vector of the ball is partially determined by the vector embedding from connectionist networks. That is, symbols are only partially landed onto the vector embedding space. Through a symbol spatialization process, we embody a symbolic structure into a continuous space, so that the brittleness problem of symbolic approaches could be solved (see Sect. 8.2.5).

We illustrate the symbol spatialization problem by the following example. Let the sentence "*it was proposed to construct a maglev train between Berlin, capital of Germany, and harbor-city Hamburg*" be in a training set, connectionist networks can

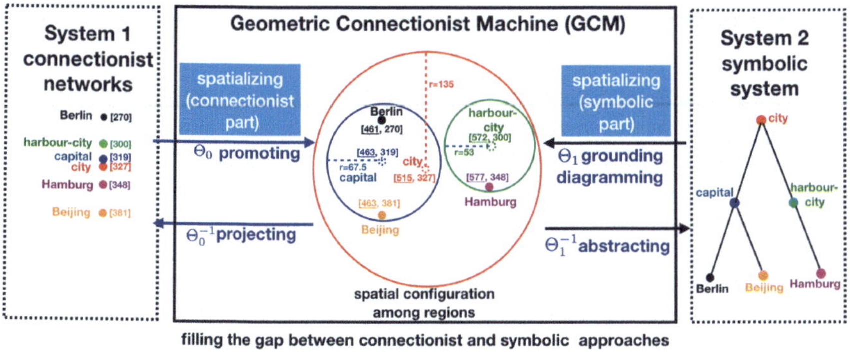

Fig. 3.9 Spatializing symbolic structures onto vector space. Connectionist network represents a word as a one-element vector. Biased by training sentence, Hamburg vector `[348]` is closer to city vector `[327]` and capital vector `[319]` than to harbor-city vector `[300]`. To precisely encode symbolic tree structures, we promote them into circles, for example, promoting harbor-city vector to a circle with central point `[572, 300]` with radius = `53`, with the aim that inclusion relations among circles encode child-parent relations in the tree structure

capture co-occurrence relations among words through word embeddings. The left-handed side of Fig. 3.9 illustrates one dimension word embeddings, such as Berlin represented by `[270]`. Suppose that Hamburg is a harbor-city, as a piece of truth-knowledge, be not directly stated in the training data. In the final word-embeddings, Hamburg vector is closer to city vector and capital vector than to harbor-city vector, which violates the symbolic structure that Hamburg is-a harbor-city and Hamburg is-not-a capital. Such biased training data restrict connectionists and statisticians to remain at the level of approximation in reasoning. Our method is to promote a one-dimensional word-embedding `[y]` into two-dimensional circles with central point `[x,y]` and radius `r`. Then, we gear all `x`s and `r`s, so that inclusion relations among circles could precisely encode the symbolic tree structure, as illustrated in the middle part of Fig. 3.9. Adding dimension is based on the assumption is that the space to embed semantic relations of words shall not be the same as the space to embed their co-occurrence relations.

The semantics of spatial relations plays the central role in human semantic systems (Regier 1997). Metaphor is grounded in spatial concepts, such as temperature, size, and orientation (Lakoff and Johnson 1980; Grady 1997). Our language and thoughts are embodied—they are rooted in the knowledge of the space around us (Feldman 2006). Spatial thinking is the foundation of abstract thought (Tversky 2019). Procedures of bottom-up learning and top-down learning (Sun 2016) can be interpreted as promoting followed by abstracting ($\Theta_0 \circ \Theta_1^{-1}$) and by grounding followed by projecting ($\Theta_1 \circ \Theta_0^{-1}$), respectively. The key is the high dimensional spatial semantic model that fills the gap between symbolic models and connectionist networks. In Chap. 8, we will see that this kind of representation resolves almost all related questions and expectations in the symbol-subsymbol debates in the literature, and in a way creates a continuum between connectionism and symbolicism.

3.6 Summary

Deep Learning simulates a number of functions of System 1 of the mind, and can be easily falsified by adversarial samples. Imposing symbolic structures onto Deep Learning systems shall enhance the robustness. A possible solution is the region-based entity representation. That is, vector embeddings, produced by Deep Learning systems, shall be promoted into balls, so that spatial relations among balls could precisely encode symbolic structures. This region-based configuration is on the way to fill the gap between continuous connectionist models and discrete symbolic structures.

References

Bordes, A., Usunier, N., Garcia-Duran, A., Weston, J., & Yakhnenko, O. (2013). Translating embeddings for modelling multi-relational data. In C. J. C. Burges, L. Bottou, M. Welling, Z. Ghahramani, & K. Q. Weinberger (Eds.), *Advances in neural information processing systems* (Vol. 26, pp. 2787–2795). Curran Associates, Inc.

Byrne, R. W. (1979). Memory for urban geography. *Quarterly Journal of Experimental Psychology, 31*, 147–154.

Dinsmore, J. (1992). Thunder in the gap. In *The symbolic and connectionist paradigms: Closing the gap* (pp. 1–23). Hillsdale, NJ: Erlbaum.

Dong, T., Wang, Z., Li, J., Bauckhage, C., & Cremers, A. B. (2019a). Triple classification using regions and fine-grained entity typing. In *Proceedings of the Thirty-Third AAAI Conference on Artificial Intelligence (AAAI-19)*, (pp. 77–85), Honolulu, Hawaii, USA, 27 January–1 February 2019.

Dong, T., Bauckhage, C., Jin, H., Li, J., Cremers, O. H., Speicher, D., Cremers, A. B., & Zimmermann, J. (2019b). Imposing category trees onto word-embeddings using a geometric construction. In *ICLR-19*, New Orleans, USA. 6–9 May 2019.

Dreyfus, H. L., Dreyfus, S. E., & Athanasiou, T. (1986). *Mind over machine: The power of human intuition and expertise in the era of the computer*. New York, NY, USA: The Free Press.

Dyer, M. G. (1988). The promise and problems of connectionism. *Behavioral and Brain Sciences, 1*, 32–33.

Erk, K. (2009). Supporting inferences in semantic space: Representing words as regions. In *IWCS-8'09* (pp. 104–115). Stroudsburg, PA, USA: Association for Computational Linguistics.

Faruqui, M., Dodge, J., Jauhar, S. K., Dyer, C., Hovy, E., & Smith, N. A. (2015). Retrofitting word vectors to semantic lexicons. In *Proceedings of the 2015 Conference of the North American Chapter of the Association for Computational Linguistics: Human Language Technologies* (pp. 1606–1615). ACL.

Feldman, J. (2006). *From molecule to metaphor: A neural theory of language*. Cambridge, MA: The MIT Press.

Fu, R., Guo, J., Qin, B., Che, W., Wang, H., & Liu, T. (2015). Learning semantic hierarchies: A continuous vector space approach. *Transactions on Audio, Speech, and Language Processing, 23*(3), 461–471.

Gärdenfors, P. (2000). *Conceptual spaces—The geometry of thought*. Cambridge, MA, USA: MIT Press.

Grady, J. (1997). *Foundations of meaning: Primary metaphors and primary scenes*, University Microfilms.

Harnad, S. (1990). The symbol grounding problem. *Physics D, 42*(1–3), 335–346.

Harnad, S. (2003). The symbol grounding problem. In *Encyclopedia of cognitive science*. Nature Publishing Group/Macmillan.
Hristov, Y., Penkov, S., Lascarides, A., & Ramamoorthy, S. (2017). Grounding symbols in multimodal instructions. In *Proceedings of the First Workshop on Language Grounding for Robotics* (pp. 49–57). Vancouver, Canada: Association for Computational Linguistics.
Ji, G., He, S., Xu, L., Liu, K., & Zhao, J. (2015). Knowledge graph embedding via dynamic mapping matrix. In *ACL'2015* (pp. 687–696). Beijing: ACL.
Kahneman, D. (2011). *Thinking, fast and slow*. Allen Lane, Penguin Books. Nobel laureate in Economics in 2002.
Lakoff, G., & Johnson, M. (1980). *Metaphors We live by*. Chicago: The University of Chicago Press. Citation is based on the reprinted in 2003.
Li, X., Vilnis, L., Zhang, D., Boratko, M., & McCallum, A. (2019). Smoothing the geometry of box embeddings. In *International Conference on Learning Representations (ICLR)*.
Lin, Y., Liu, Z., Sun, M., Liu, Y., & Zhu, X. (2015). Learning entity and relation embeddings for knowledge graph completion. In *AAAI'15* (pp. 2181–2187). AAAI Press.
McNamara, T. P. (1991). Memory's view of space. *The Psychology of Learning and Motivation*, *27*, 147–186.
Quillian, M. (1968). Semantic memory. In M. Minsky (Ed.), *Semantic information processing*. Cambridge, MA: MIT Press.
Regier, T. (1997). *The human semantic potential: Spatial language and constrained connectionism*. Cambridge, MA: The MIT Press.
Steels, L. (2008). The symbol grounding problem has been solved. So what's next. In *Symbols and embodiment: Debates on meaning and cognition* (pp. 223–244). New Orleans, USA: Oxford University Press.
Stevens, A., & Coupe, P. (1978). Distance estimation from cognitive maps. *Cognitive Psychology*, *13*, 526–550.
Sun, R. (2015). Artificial intelligence: Connectionist and symbolic approaches. In D. W. James (Ed.), *International encyclopedia of the social and behavioral sciences* (2nd ed., pp. 35–40). Oxford: Pergamon/Elsevier.
Sun, R. (2016). Implicit and explicit processes: Their relation, interaction, and competition. In L. Macchi, M. Bagassi, & R. Viale (Eds.), *Cognitive unconscious and human rationality* (pp. 27–257). Cambridge, MA: MIT Press.
Tversky, B. (1981). Distortions in memory for maps. *Cognitive Psychology*, *13*, 407–433.
Tversky, B. (1992). Distortions in cognitive maps. *Geoforum*, *23*(2), 131–138.
Tversky, B. (2019). *Mind in motion*. New York, USA: Basic Books.
Wang, Z., Zhang, J., Feng, J., & Chen, Z. (2014). Knowledge graph embedding by translating on hyperplanes. In *AAAI* (pp. 1112–1119). AAAI Press.

Chapter 4
The Criteria, Challenges, and the Back-Propagation Method

In this chapter, we describe our task of symbol spatialization, list the criteria and challenges. We show that despite of the magic power, back-propagation method will not be the right tool to fulfill the criteria.

4.1 Spatializing Symbolic Tree Structures Onto Vector Embeddings

The tree structure is one of the fundamental data structures in computer science, and was targeted by connectionists to demonstrate the representation power of connectionist networks (Pollack 1990; Adamson and Damper 1999). Words in similar contexts have similar semantic and syntactic information. Word embeddings are vector representations of words that reflect this characteristic (Mikolov et al. 2013; Pennington et al. 2014), and have been widely used in AI applications, such as question-answering (Tellex et al. 2003), text classification (Sebastiani 2002), information retrieval (Manning et al. 2008), or even as a building-block for a unified NLP system to process common NLP tasks (Collobert et al. 2011). A promising research direction to enhance semantic reasoning is to extract semantic relations from corpus and explicitly represent semantic relations in the embedding space (Erk 2009; Lenci and Benotto 2012; Kruszewski et al. 2015; Fu et al. 2015; Nickel and Kiela 2017; Li et al. 2019). One widely used semantic database is the WordNet (Miller 1995). Our task is to take two kinds of different information as input: one is the pre-trained GloVe word-embeddings (Pennington et al. 2014), the other is symbolic tree structures extracted from WordNet 3.0, and precisely spatialize tree structures onto word-embeddings.

T. Dong, *A Geometric Approach to the Unification of Symbolic Structures and Neural Networks*, Studies in Computational Intelligence 910,
https://doi.org/10.1007/978-3-030-56275-5_4

4.2 The Challenge of Zero Energy Cost

As discussed in Sect. 3.4, we set out the strict criteria as follows: (1) all relations in the tree structure must be correctly spatialized. Precisely, every symbolic child-parent relation shall be encoded by the inclusion relation between the child ball and the parent ball; sibling nodes shall be encoded by mutually disconnected sibling balls; (2) The given vector embeddings shall be well-preserved. Extending vector embeddings onto region embeddings is not a new idea in the literature (Athiwaratkun and Wilson 2017; Xiao et al. 2016; Nickel and Kiela 2017), however, our strict criteria is a challenging task that has not been achieved, even not yet been targeted in the literature of representation learning.

The challenge lies in the fact that in many cases, words, such as ice_cream, tuberose, and their superordinate words, such as dessert, plant, seldom occur in the same context, their word embeddings differ to such a degree that the cosine value is less than zero. For example, using GloVe embedding (Pennington et al. 2014), the cos value of ice_cream and dessert is −0.1998, the cos value of tuberose and plant is −0.2191. This follows that dessert ball, if it contains ice_cream ball, will contain the origin point of the embedding space. Plant ball, if it contains tuberose ball, will contains the origin point of the embedding space. Then, dessert ball overlaps with plant ball, as illustrated in Fig. 4.1. As there is no entity which can be subordinated to both dessert and plant, dessert ball shall disconnect from plant ball. Given a large knowledge tree, how could we guarantee this?

In the literature of word-embedding and graph-embedding, the termination condition of back-propagation training processes is a local minimum. This local minimum is greater than zero. As we need to precisely encode all symbolic relations into inclusion relations among regions, we require zero energy loss (the global minimum). We call this *the Challenge of Zero Energy Cost.*

Can back-propagation method (Rumelhart et al. 1986; LeCun et al. 2015) fulfill our criteria? We utilize back-propagation method to solve two tasks: one is the way-finding task in a maze, the other is a task of spatializing small scaled trees. The first task is satisfyingly approached, while the second task not. We end up with the conclusion that the back-propagation method cannot guarantee zero energy cost for the symbol spatialization task.

4.3 The Back-Propagation Method for Way-Finding

The back-propagation method is the fundamental method for connectionist approaches (Energy-Based Model) (LeCun et al. 2006). We revisit this method by applying it for predicting next route instructions in a physical space, which is a basic task of cognition (Tversky 2019). In Sect. 6.5.4, we will interpret the task of membership validation as a task of way-finding in embedding spaces Fig. 4.2.

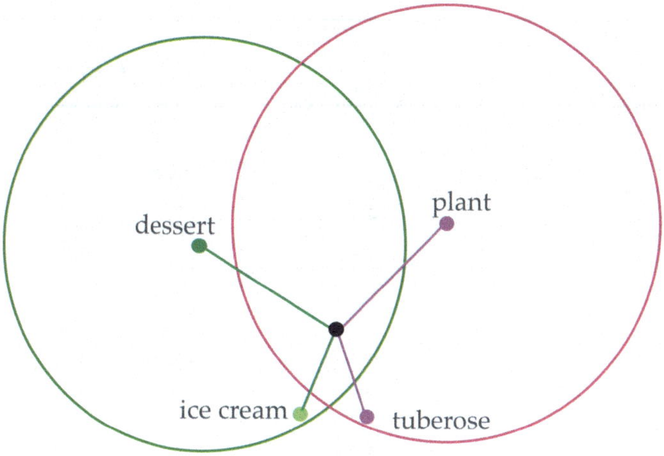

Fig. 4.1 Dessert ball partially overlaps with plant ball, although they should be disconnected from each other

Fig. 4.2 You are in this maze, equipped with a computer, and given a partial route instruction. Now, the instruction reaches the end, you are still in the maze, which direction shall you take in the next crossing point?

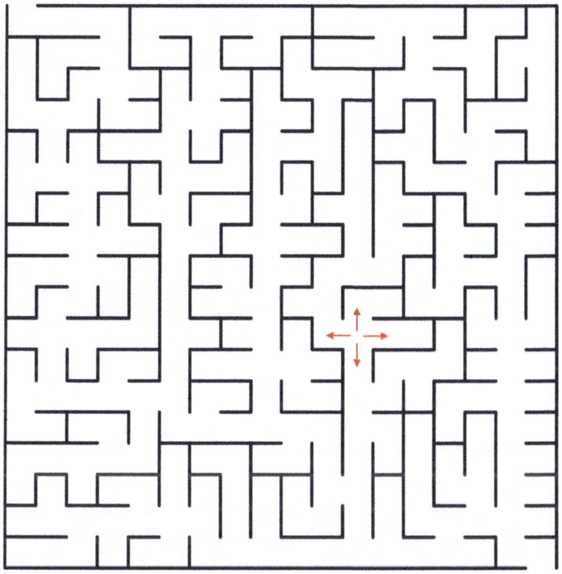

Suppose that you are put in a large maze, given a long route instruction that only consists of four unit instructions, *turn-left, turn-right, go-ahead, turn-back*, to inform you the action in the next decision point. You loyally followed the instructions, and unfortunately, you are still in the maze, when the instruction list goes to the end. What can you do? Build a connectionist network to tell you the next action.

Let $ROUTE$ be the given route instruction $[R_1, R_2, \ldots, R_n]$. As there are only four unit instructions, we encode R_i by a two-element vector as follows:

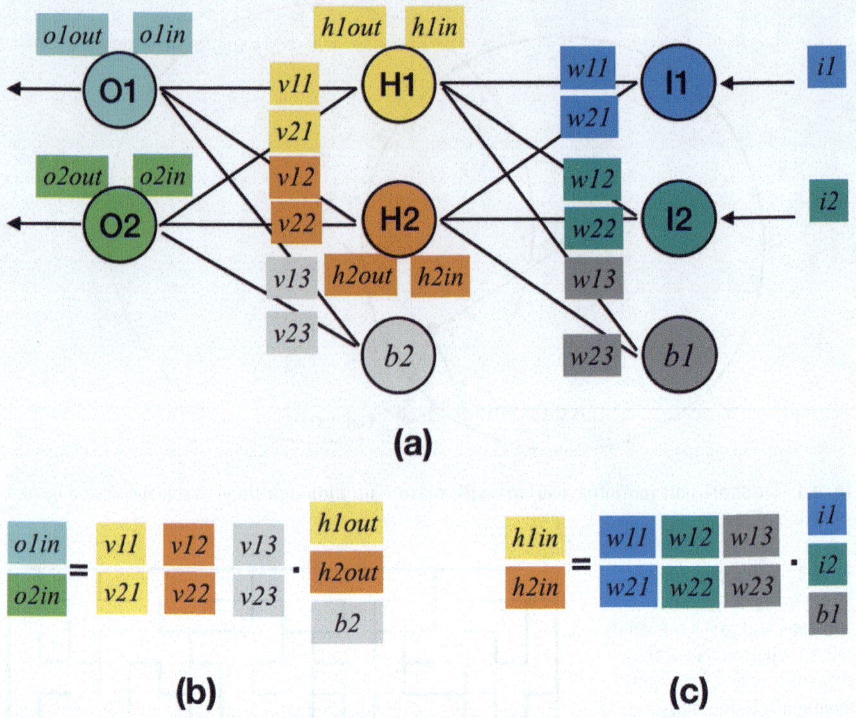

Fig. 4.3 **a** A connectionist network with one hidden layer, two input nodes, and two output nodes; **b** matrix computation of the second layer; **c** matrix computation of the first layer

$$R_1 : \textit{turn-left} \triangleq \begin{bmatrix} 1 \\ 0 \end{bmatrix}, R_2 : \textit{turn-right} \triangleq \begin{bmatrix} -1 \\ 0 \end{bmatrix}$$
$$R_3 : \textit{go-ahead} \triangleq \begin{bmatrix} 0 \\ 1 \end{bmatrix}, R_4 : \textit{turn-around} \triangleq \begin{bmatrix} 0 \\ -1 \end{bmatrix}$$

We design a simple connectionist network with one hidden layer, two input nodes, two output nodes, and two bias nodes b_1 and b_2, as illustrated in Fig. 4.3a. The intended function of this network is to predict the next route instruction, given the current route instruction. Our work is to gear the 12 parameters of this network: $w_{11}, \ldots, w_{32}, v_{11}, \ldots, v_{32}$ (we set $b_1 = b_2 = 1$). We initialize them randomly, and set R_1 as the input, to see whether we are lucky enough to have R_2 as output (normally, we do not). We update these parameters to improve the quality of the network. Let us go through the detail.

4.3.1 Matrix Computing

The inputs of the hidden layer are computed by the following matrix, see Fig. 4.3c

$$\begin{bmatrix} h_{1in} \\ h_{2in} \end{bmatrix} = \begin{bmatrix} w_{11} & w_{12} & w_{13} \\ w_{21} & w_{22} & w_{23} \end{bmatrix} \cdot \begin{bmatrix} i_1 \\ i_2 \\ b_1 \end{bmatrix} \tag{4.1}$$

The logistic function used in the hidden layer are as follows.

$$h_{1out} = \frac{1}{1+e^{-h_{1in}}} \tag{4.2}$$

$$h_{2out} = \frac{1}{1+e^{-h_{2in}}} \tag{4.3}$$

The inputs to the output layer are computed by the following matrix, see Fig. 4.3b

$$\begin{bmatrix} o_{1in} \\ o_{2in} \end{bmatrix} = \begin{bmatrix} v_{11} & v_{12} & v_{13} \\ v_{21} & v_{22} & v_{23} \end{bmatrix} \cdot \begin{bmatrix} h_{1out} \\ h_{2out} \\ b_2 \end{bmatrix} \tag{4.4}$$

The logistic function used in the output layer are as follows.

$$o_{1out} = \frac{2}{1+e^{-o_{1in}}} - 1 \tag{4.5}$$

$$o_{2out} = \frac{2}{1+e^{-o_{2in}}} - 1 \tag{4.6}$$

4.3.2 Forward Computing

We initialize $w_{ij} = v_{ij} = 0.5$, and $b_1 = b_2 = 1$. The first instruction in the route is R_1.

$$\begin{bmatrix} h_{1in} \\ h_{2in} \end{bmatrix} = \begin{bmatrix} w_{11} & w_{12} & w_{13} \\ w_{21} & w_{22} & w_{23} \end{bmatrix} \cdot \begin{bmatrix} i_1 \\ i_2 \\ b_1 \end{bmatrix} = \begin{bmatrix} 0.5 & 0.5 & 0.5 \\ 0.5 & 0.5 & 0.5 \end{bmatrix} \cdot \begin{bmatrix} 1 \\ 0 \\ 1 \end{bmatrix} = \begin{bmatrix} 1 \\ 1 \end{bmatrix} \tag{4.7}$$

$$h_{1out} = \frac{1}{1+e^{-h_{1in}}} = \frac{1}{1+e^{-1}} = 0.731 \tag{4.8}$$

$$h_{2out} = \frac{1}{1+e^{-h_{2in}}} = \frac{1}{1+e^{-1}} = 0.731 \tag{4.9}$$

$$\begin{bmatrix} o_{1in} \\ o_{2in} \end{bmatrix} = \begin{bmatrix} v_{11} & v_{12} & v_{13} \\ v_{21} & v_{22} & v_{23} \end{bmatrix} \cdot \begin{bmatrix} h_{1out} \\ h_{2out} \\ b_2 \end{bmatrix} = \begin{bmatrix} 0.5 & 0.5 & 0.5 \\ 0.5 & 0.5 & 0.5 \end{bmatrix} \cdot \begin{bmatrix} 0.731 \\ 0.731 \\ 1 \end{bmatrix} = \begin{bmatrix} 1.231 \\ 1.231 \end{bmatrix} \tag{4.10}$$

$$o_{1out} = \frac{2}{1+e^{-o_{1in}}} - 1 = \frac{2}{1+e^{-1.231}} - 1 = 0.547 \tag{4.11}$$

$$o_{2out} = \frac{2}{1+e^{-o_{2in}}} - 1 = \frac{2}{1+e^{-1.231}} - 1 = 0.547 \tag{4.12}$$

We would interpret the result as a route instruction, which is an orientation information. The second instruction should be $R_2 = \begin{bmatrix} -1 \\ 0 \end{bmatrix}$, which is quite different to the predicted value. We use $\cos(R_{true}, R_{predicted})$ to measure the quality of the predicted orientation: the best case would be 1, the worse case would be -1. We compute the error

$$E = 1 - \cos(R_{true}, R_{predicted})$$

so that in the best case $E = 0$, and in the worse case $E = 2$. Our first error value $E = 1 - \cos\left(\begin{bmatrix} -1 \\ 0 \end{bmatrix}, \begin{bmatrix} 0.547 \\ 0.547 \end{bmatrix}\right) = 1.7071$.

4.3.3 Backward Updating

We need to update parameters to reduce errors. Using back-propagation algorithm, we only need to compute the partial derivative of each parameter to the total error E.

$$\frac{\partial E}{\partial v_{11}} = \frac{\partial E}{\partial o_{1out}} \times \frac{\partial o_{1out}}{\partial o_{1in}} \times \frac{\partial o_{1in}}{\partial v_{11}} \tag{4.13}$$

$$E = 1 - \cos\left(\begin{bmatrix} o_{1true} \\ o_{2true} \end{bmatrix}, \begin{bmatrix} o_{1out} \\ o_{2out} \end{bmatrix}\right) = 1 - \frac{o_{1true}o_{1out} + o_{2true}o_{2out}}{\sqrt{o_{1true}^2 + o_{2true}^2}\sqrt{o_{1out}^2 + o_{2out}^2}} \tag{4.14}$$

$$\frac{\partial E}{\partial o_{1out}} = -\frac{\partial \frac{o_{1true}o_{1out}+o_{2true}o_{2out}}{\sqrt{o_{1true}^2+o_{2true}^2}\sqrt{o_{1out}^2+o_{2out}^2}}}{\partial o_{1out}} \tag{4.15}$$

$$= -\frac{\sqrt{o_{1true}^2 + o_{2true}^2}\sqrt{o_{1out}^2 + o_{2out}^2}\frac{\partial o_{1true}o_{1out}+o_{2true}o_{2out}}{\partial o_{1out}}}{(o_{1true}^2 + o_{2true}^2)(o_{1out}^2 + o_{2out}^2)} \tag{4.16}$$

$$+\frac{(o_{1true}o_{1out} + o_{2true}o_{2out})\frac{\partial \sqrt{o_{1true}^2+o_{2true}^2}\sqrt{o_{1out}^2+o_{2out}^2}}{\partial o_{1out}}}{(o_{1true}^2 + o_{2true}^2)(o_{1out}^2 + o_{2out}^2)} \tag{4.17}$$

$$= -\frac{o_{1true}}{\sqrt{o_{1true}^2 + o_{2true}^2}\sqrt{o_{1out}^2 + o_{2out}^2}} \tag{4.18}$$

$$+\frac{(o_{1true}o_{1out} + o_{2true}o_{2out})o_{1out}}{(o_{1out}^2 + o_{2out}^2)^{\frac{3}{2}}\sqrt{o_{1true}^2 + o_{2true}^2}} \tag{4.19}$$

$$= -\frac{-1}{\sqrt{(-1)^2 + 0^2}\sqrt{0.547^2 + 0.547^2}} \tag{4.20}$$

$$+\frac{(-1 \times 0.547 + 0 \times 0.547) \times 0.547}{(0.547^2 + 0.547^2)^{\frac{3}{2}}\sqrt{(-1)^2 + 0^2}} \tag{4.21}$$

$$= 0.6463 \tag{4.22}$$

$$\frac{\partial o_{1out}}{\partial o_{1in}} = o_{1out}(1 - o_{1out}) = 0.547 \times (1 - 0.547) = 0.2478 \tag{4.23}$$

$$\frac{\partial o_{1in}}{\partial v_{11}} = \frac{\partial(v_{11}h_{1out} + v_{12}h_{2out} + v_{13}b_2)}{\partial v_{11}} = h_{1out} = 0.731 \tag{4.24}$$

Therefore,

$$\frac{\partial E}{\partial v_{11}} = \frac{\partial E}{\partial o_{1out}} \times \frac{\partial o_{1out}}{\partial o_{1in}} \times \frac{\partial o_{1in}}{\partial v_{11}} \tag{4.25}$$

$$= 0.6463 \times 0.2478 \times 0.731 \tag{4.26}$$

$$= 0.1171 \tag{4.27}$$

To decrease the value of E, we need to decrease v_{11} as follows.

$$v_{11}^{(next)} = v_{11} - \eta\frac{\partial E}{\partial v_{11}} \tag{4.28}$$

$$= 0.5 - 10 \times 0.1171 \tag{4.29}$$

$$= -0.671 \tag{4.30}$$

η is the learning rate, normally is set around 0.001.

In the same way, we have

$$v_{12}^{(next)} = v_{12} - \eta\frac{\partial E}{\partial v_{12}} = 0.5 - 10 \times 0.1171 = -0.671 \tag{4.31}$$

$$v_{13}^{(next)} = v_{13} - \eta\frac{\partial E}{\partial v_{13}} = 0.5 - 10 \times 0.160 = -1.1 \tag{4.32}$$

$$v_{21}^{(next)} = v_{21} - \eta\frac{\partial E}{\partial v_{21}} = 0.5 - 10 \times (-0.1171) = 1.6171 \tag{4.33}$$

$$v_{22}^{(next)} = v_{22} - \eta\frac{\partial E}{\partial v_{22}} = 0.5 - 10 \times (-0.1171) = 1.6171 \tag{4.34}$$

$$v_{23}^{(next)} = v_{23} - \eta \frac{\partial E}{\partial v_{23}} = 0.5 - 10 \times (-0.160) = 2.1 \tag{4.35}$$

Next, we need to update w_{ij}.

$$\frac{\partial E}{\partial w_{11}} = \frac{\partial E}{\partial h_{1out}} \times \frac{\partial h_{1out}}{\partial h_{1in}} \times \frac{\partial h_{1in}}{\partial w_{11}} \tag{4.36}$$

$$= \frac{\partial E}{\partial o_{1out}} \times \frac{\partial o_{1out}}{\partial o_{1in}} \times \frac{\partial o_{1in}}{\partial h_{1out}} \times \frac{\partial h_{1out}}{\partial h_{1in}} \times \frac{\partial h_{1in}}{\partial w_{11}} \tag{4.37}$$

$$+ \frac{\partial E}{\partial o_{2out}} \times \frac{\partial o_{2out}}{\partial o_{2in}} \times \frac{\partial o_{2in}}{\partial h_{1out}} \times \frac{\partial h_{1out}}{\partial h_{1in}} \times \frac{\partial h_{1in}}{\partial w_{11}} \tag{4.38}$$

$$= \frac{\partial E}{\partial o_{1out}} \times \frac{\partial o_{1out}}{\partial o_{1in}} v_{11}^{(next)} \times \frac{\partial h_{1out}}{\partial h_{1in}} \times \frac{\partial h_{1in}}{\partial w_{11}} \tag{4.39}$$

$$+ \frac{\partial E}{\partial o_{2out}} \times \frac{\partial o_{2out}}{\partial o_{2in}} v_{21}^{(next)} \times \frac{\partial h_{1out}}{\partial h_{1in}} \times \frac{\partial h_{1in}}{\partial w_{11}} \tag{4.40}$$

$$= 0.6463 \times 0.2478 \times (-0.671) \times 0.731 \times 0.269 \times 1 \tag{4.41}$$

$$+ (-0.6463) \times 0.2478 \times 1.6171 \times 0.731 \times 0.269 \times 1 \tag{4.42}$$

$$= -0.0721 \tag{4.43}$$

$$w_{11}^{(next)} = w_{11} - \eta \frac{\partial E}{\partial w_{11}} = 0.5 - 10 \times (-0.0721) = 1.221 \tag{4.44}$$

$$w_{12}^{(next)} = w_{12} - \eta \frac{\partial E}{\partial w_{12}} = 0.5 - 10 \times 0 = 0 \tag{4.45}$$

$$w_{13}^{(next)} = w_{13} - \eta \frac{\partial E}{\partial w_{13}} = 0.5 - 10 \times (-0.0721) = 1.221 \tag{4.46}$$

$$w_{21}^{(next)} = w_{21} - \eta \frac{\partial E}{\partial w_{21}} = 0.5 - 10 \times -0.0721 = 1.221 \tag{4.47}$$

$$w_{22}^{(next)} = w_{22} - \eta \frac{\partial E}{\partial w_{22}} = 0.5 - 10 \times 0 = 0 \tag{4.48}$$

$$w_{23}^{(next)} = w_{23} - \eta \frac{\partial E}{\partial w_{23}} = 0.5 - 10 \times -0.0721 = 1.221 \tag{4.49}$$

After these parameters are updated, we have the

$$\begin{bmatrix} h_{1in} \\ h_{2in} \end{bmatrix} = \begin{bmatrix} w_{11} \; w_{12} \; w_{13} \\ w_{21} \; w_{22} \; w_{23} \end{bmatrix} \cdot \begin{bmatrix} i_1 \\ i_2 \\ b_1 \end{bmatrix} = \begin{bmatrix} 1.221 \; 0 \; 1.221 \\ 1.221 \; 0 \; 1.221 \end{bmatrix} \cdot \begin{bmatrix} 1 \\ 0 \\ 1 \end{bmatrix} = \begin{bmatrix} 2.442 \\ 2.442 \end{bmatrix} \tag{4.50}$$

$$h_{1out} = \frac{1}{1 + e^{-h_{1in}}} = \frac{1}{1 + e^{-2.442}} = 0.92 \tag{4.51}$$

$$h_{2out} = \frac{1}{1 + e^{-h_{2in}}} = \frac{1}{1 + e^{-2.442}} = 0.92 \tag{4.52}$$

$$\begin{bmatrix} o_{1in} \\ o_{2in} \end{bmatrix} = \begin{bmatrix} v_{11} & v_{12} & v_{13} \\ v_{21} & v_{22} & v_{23} \end{bmatrix} \cdot \begin{bmatrix} h_{1out} \\ h_{2out} \\ b_2 \end{bmatrix} = \begin{bmatrix} -0.671 & -0.671 & -1.1 \\ 1.6171 & 1.6171 & 2.1 \end{bmatrix} \cdot \begin{bmatrix} 0.92 \\ 0.92 \\ 1 \end{bmatrix} \tag{4.53}$$

$$= \begin{bmatrix} -2.335 \\ 5.075 \end{bmatrix} \tag{4.54}$$

$$o_{1out} = \frac{2}{1+e^{-o_{1in}}} - 1 = \frac{2}{1+e^{2.335}} - 1 = -0.823 \tag{4.55}$$

$$o_{2out} = \frac{2}{1+e^{-o_{2in}}} - 1 = \frac{2}{1+e^{-5.075}} - 1 = 0.988 \tag{4.56}$$

With these updated parameters, the error E reduces to $1 - \cos\left(\begin{bmatrix} -1 \\ 0 \end{bmatrix}, \begin{bmatrix} -0.823 \\ 0.988 \end{bmatrix}\right) = 0.35$. Back propagation method is amazing. It works actually very well. It points out the almost right direction, and may save your life. Would you use navigators using this method? You are very likely to say "No". But, what if the number of directions increases from 4 to millions? What if the manufacturer tells you that the navigator is learned by route instructions of millions of mazes in the world? Then you will use it. Probably, you already have one—Word-Embedding (Mikolov 2012; Mikolov et al. 2013), in which the same back propagation method tells you what should be the next word (among millions of candidate words) of the current sentence segment, after learning from billions of sentences. You might be even manufacture similar products using this back-propagation method—many others have tried and succeeded in a variety of tasks (Bengio et al. 2003; Mikolov 2012; Mikolov et al. 2013; Socher et al. 2013; Faruqui et al. 2015; Xiao et al. 2016; Han et al. 2016; Speer et al. 2017; LeCun et al. 2015). Could the back propagation method continue to be successful for the symbol spatialization task? This task appears deceptively simple. We only need to add two new elements for each vector, one representing the length of the central point vector, the other representing the radius, and use back-propagation method to gear the two kinds of new parameters.

4.4 Back-Propagation for Symbol Spatialization

In this section, we present our experiment on a toy system to show that back-propagation method cannot guarantee to guarantee zero energy cost in the task of symbol spatialization.

4.4.1 A Toy System to Test Back-Propagation Method

We design a small knowledge-graph consisting of two tree structures totaling 10 relations, as illustrated in Fig. 4.4 and listed in Table 4.1. Their initial states are listed in Table 4.2 and illustrated in Fig. 4.5.

4.4.2 Updating Configuration Using Back-Propagation Method

A ball is structured by a radius r ($r > 0$) and a central point vector $\mathbf{O}$, $\mathbf{O} = (\boldsymbol{\alpha}, l)$, in which $\boldsymbol{\alpha}$ and l are the direction and the length of $\mathbf{O}$, respectively. We define Ball $\mathbb{B}$ as the set of point $\mathbf{p}$ satisfying $\|l \cdot \boldsymbol{\alpha} - \mathbf{p}\| < r$, written as

$$\mathbb{B}(\mathbf{O}, r) = \mathbb{B}((\boldsymbol{\alpha}, l), r)$$

as shown in Fig. 4.6.

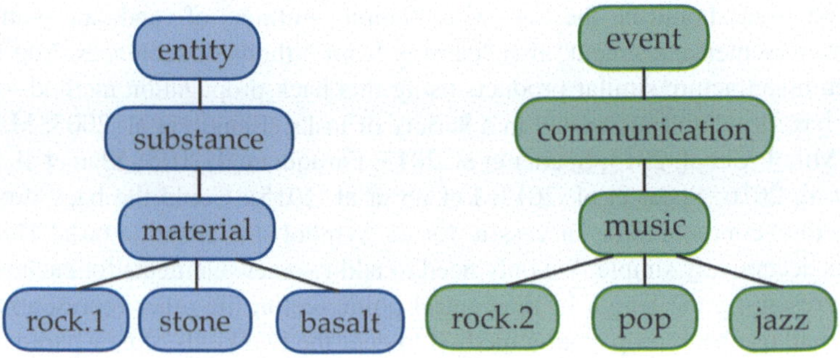

Fig. 4.4 A small knowledge-base with two tree structures

Table 4.1 Two tree structures in the small knowledge-base

Two tree structures					
Child	Relation	Parent	Child	Relation	Parent
rock.1	is_a	Material	rock.2	is_a	Music
Stone	is_a	Material	Pop	is_a	Music
Basalt	is_a	Material	Jazz	is_a	Music
Material	Subordinate	Substance	Music	Subordinate	Communication
Substance	Subordinate	Entity	Communication	Subordinate	Event

Table 4.2 Each entity is represented by a 2-dimensional ball

Initial location and size of $\mathcal{N}$-Balls				
Name	The x orientation of the central vector	The y orientation of the central vector	The length of the central vector	The radius
rock.1	$\cos(\frac{\pi}{4})$	$\sin(\frac{\pi}{4})$	10	0.05
Stone	$\cos(\frac{\pi}{4}+\frac{\pi}{100})$	$\sin(\frac{\pi}{4}+\frac{\pi}{100})$	10	0.05
Basalt	$\cos(\frac{\pi}{4}-\frac{\pi}{200})$	$\sin(\frac{\pi}{4}-\frac{\pi}{200})$	10	0.05
Material	$\cos(\frac{\pi}{4}-\frac{\pi}{100})$	$\sin(\frac{\pi}{4}-\frac{\pi}{100})$	10	0.05
Substance	$\cos(\frac{\pi}{4}-\frac{\pi}{50})$	$\sin(\frac{\pi}{4}-\frac{\pi}{50})$	10	0.05
Entity	$\cos(\frac{\pi}{4}-\frac{\pi}{40})$	$\sin(\frac{\pi}{4}-\frac{\pi}{40})$	10	0.05
rock.2	$\cos(\frac{\pi}{4}+\frac{\pi}{30})$	$\sin(\frac{\pi}{4}+\frac{\pi}{30})$	10	0.05
Pop	$\cos(\frac{\pi}{4}+\frac{\pi}{50})$	$\sin(\frac{\pi}{4}+\frac{\pi}{50})$	10	0.05
Jazz	$\cos(\frac{\pi}{4}-\frac{\pi}{50})$	$\sin(\frac{\pi}{4}-\frac{\pi}{50})$	10	0.05
Music	$\cos(\frac{\pi}{3}+\frac{\pi}{15})$	$\sin(\frac{\pi}{3}+\frac{\pi}{15}))$	10	0.05
Communication	$\cos(\frac{\pi}{3})$	$\sin(\frac{\pi}{3})$	10	0.05
Event	$\cos(\frac{\pi}{3}-\frac{\pi}{50})$	$\sin(\frac{\pi}{3}+\frac{\pi}{50})$	10	0.05

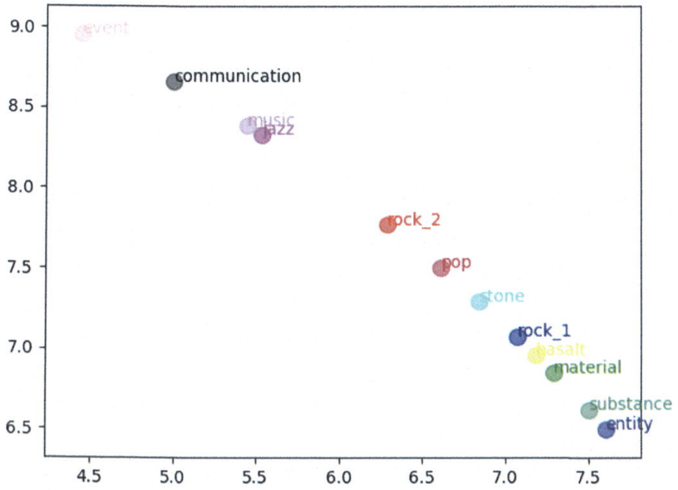

Fig. 4.5 Diagrammatic representation of the initial entity balls

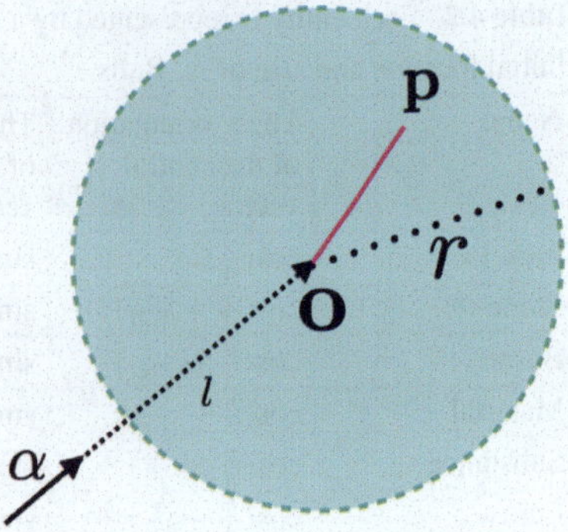

Fig. 4.6 The structure of a ball in 2-dimensional space is defined as an open region

4.4.2.1 Spatial Relations Between Balls

Ball $\mathbb{B}_1$ being inside Ball $\mathbb{B}_2$, if and only if the radius of $\mathbb{B}_2$ minus the radius of $\mathbb{B}_1$ is greater than or equal to the distance between their central points, written as

$$\mathbf{P}(\mathbb{B}_1, \mathbb{B}_2) \triangleq r_2 - r_1 \geq \|l_2 \cdot \boldsymbol{\alpha_2} - l_1 \cdot \boldsymbol{\alpha_1}\|$$

We introduce $f_{\mathbf{P}}(\mathbb{B}_1, \mathbb{B}_2) \triangleq \|l_2 \cdot \boldsymbol{\alpha_2} - l_1 \cdot \boldsymbol{\alpha_1}\| + r_1 - r_2$. $f_{\mathbf{P}}(\mathbb{B}_1, \mathbb{B}_2) \leq 0$, if and only if $\mathbb{B}_1$ is inside $\mathbb{B}_2$, as illustrated in Fig. 4.7.

Being inside is a transitive relation. If Ball $\mathbb{B}_1$ is inside Ball $\mathbb{B}_2$, and Ball $\mathbb{B}_2$ is inside Ball $\mathbb{B}_3$, then Ball $\mathbb{B}_1$ is inside Ball $\mathbb{B}_3$, as shown below.

$$\begin{aligned} f_{\mathbf{P}}(\mathbb{B}_1, \mathbb{B}_3) &\triangleq \|l_3 \cdot \boldsymbol{\alpha_3} | l_1 \cdot \boldsymbol{\alpha_1}\| + r_1 - r_3 \\ &= \|l_3 \cdot \boldsymbol{\alpha_3} | l_1 \cdot \boldsymbol{\alpha_1}\| + r_1 - r_3 + \\ &\quad \|l_2 \cdot \boldsymbol{\alpha_2} | l_1 \cdot \boldsymbol{\alpha_1}\| + r_1 - r_2 - \|l_2 \cdot \boldsymbol{\alpha_2} - l_1 \cdot \boldsymbol{\alpha_1}\| - r_1 + \\ &\quad \|l_3 \cdot \boldsymbol{\alpha_3} - l_2 \cdot \boldsymbol{\alpha_2}\| + r_2 - r_3 - \|l_3 \cdot \boldsymbol{\alpha_3} - l_2 \cdot \boldsymbol{\alpha_2}\| + r_3 \\ &= f_{\mathbf{P}}(\mathbb{B}_1, \mathbb{B}_2) + f_{\mathbf{P}}(\mathbb{B}_2, \mathbb{B}_3) + \end{aligned}$$

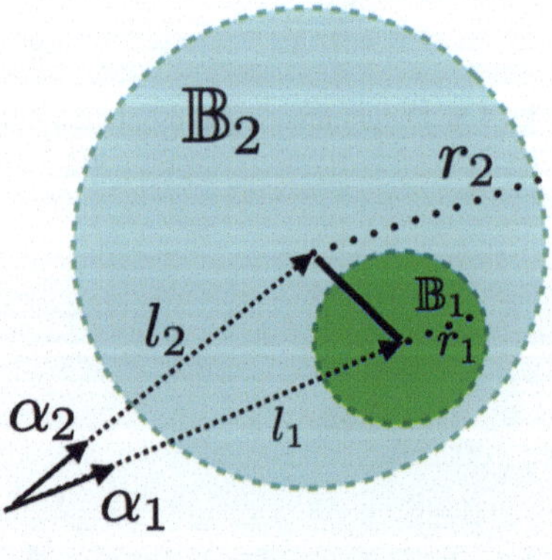

Fig. 4.7 Ball $\mathbb{B}_1$ is inside Ball $\mathbb{B}_2$

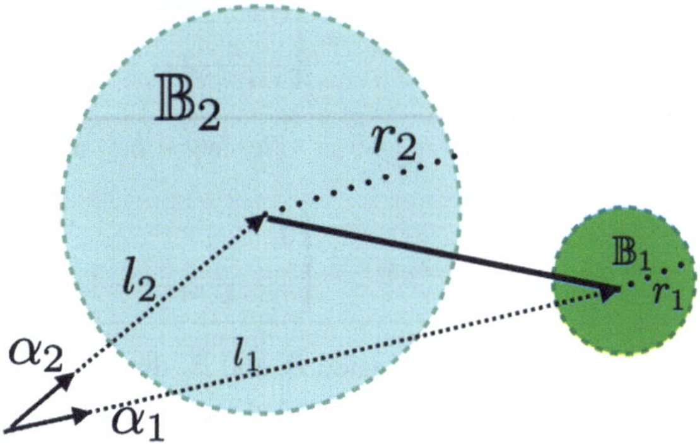

Fig. 4.8 Ball $\mathbb{B}_1$ disconnects from $\mathbb{B}_2$

$$\begin{aligned}&\|l_3 \cdot \boldsymbol{\alpha_3} - l_1 \cdot \boldsymbol{\alpha_1}\| - \|l_2 \cdot \boldsymbol{\alpha_2} - l_1 \cdot \boldsymbol{\alpha_1}\| - \|l_3 \cdot \boldsymbol{\alpha_3} - l_2 \cdot \boldsymbol{\alpha_2}\| \\ &\leq f_{\mathbf{P}}(\mathbb{B}_1, \mathbb{B}_2) \\ &\leq 0\end{aligned}$$

From the above formula, we conclude that the parent of $\mathbb{B}_1$ is the ball that produces the maximum value of $f_{\mathbf{P}}(\mathbb{B}_1, \mathbb{B}_i) \leq 0$, in which $\mathbb{B}_i$ are ancestors of $\mathbb{B}_1$. We can define the parent and sibling relations as follows.

$$\mathbf{PARENT}(\mathbb{B}_1) = \arg\max_{\mathbb{B}_i} f_{\mathbf{P}}(\mathbb{B}_1, \mathbb{B}_i) \leq 0$$

$$\mathbf{SIBLING}(\mathbb{B}_i, \mathbb{B}_j) \triangleq \mathbf{PARENT}(\mathbb{B}_i) = \mathbf{PARENT}(\mathbb{B}_j)$$

Ball $\mathbb{B}_1$ disconnects from Ball $\mathbb{B}_2$, if and only if the sum of their radii is less than or equal to the distance between their central points, written as

$$\mathbf{DC}(\mathbb{B}_0, \mathbb{B}) \stackrel{\text{def}}{=} r_{\mathbb{B}} + r_{\mathbb{B}_0} \leq \|l_{\mathbb{B}} \cdot \boldsymbol{\alpha}_{\mathbb{B}} - l_{\mathbb{B}_0} \cdot \boldsymbol{\alpha}_{\mathbb{B}_0}\|$$

We define $f_{\mathbf{DC}}(\mathbb{B}_0, \mathbb{B}) = r_{\mathbb{B}_0} + r_{\mathbb{B}} - \|l_{\mathbb{B}} \cdot \boldsymbol{\alpha}_{\mathbb{B}} - l_{\mathbb{B}_0} \cdot \boldsymbol{\alpha}_{\mathbb{B}_0}\|$. $f_{\mathbf{DC}}(\mathbb{B}_0, \mathbb{B}) \leq 0$, if and only if $\mathbb{B}_1$ disconnects from $\mathbb{B}_2$, as illustrated in Fig. 4.8.

Interestingly, when $r_{\mathbb{B}} + r_{\mathbb{B}_0} = \|l_{\mathbb{B}} \cdot \boldsymbol{\alpha}_{\mathbb{B}} - l_{\mathbb{B}_0} \cdot \boldsymbol{\alpha}_{\mathbb{B}_0}\|$, Ball $\mathbb{B}_1$ and Ball $\mathbb{B}_2$ are defined as *being disconnected*. This is consistent to the understanding of the connectedness in General Topology (Kelley 1955), in which *being connected* is defined as one region and the closure of the other region share a common point.

4.4.2.2 Energy Function and Loss Function

For $f_{\mathbf{P}}(\mathbb{B}_1, \mathbb{B}_2) \leq 0$, if and only if $\mathbb{B}_1$ is inside $\mathbb{B}_2$, we define Energy Function $E_{f_{\mathbf{P}}(\mathbb{B}_1,\mathbb{B}_2)}$ as follows.

$$E_{f_{\mathbf{P}}(\mathbb{B}_1,\mathbb{B}_2)} = \begin{cases} 0, & f_{\mathbf{P}}(\mathbb{B}_1,\mathbb{B}_2) \le 0 \\ \frac{2}{1+e^{-f_{\mathbf{P}}(\mathbb{B}_1,\mathbb{B}_2)}} - 1, & f_{\mathbf{P}}(\mathbb{B}_1,\mathbb{B}_2) > 0 \end{cases}$$

For $f_{\mathbf{DC}}(\mathbb{B}_1,\mathbb{B}_2) = r_{\mathbb{B}_0} + r_{\mathbb{B}} - \|l_{\mathbb{B}} \cdot \boldsymbol{\alpha}_{\mathbb{B}} - l_{\mathbb{B}_0} \cdot \boldsymbol{\alpha}_{\mathbb{B}_0}\|$, we define Energy Function $E_{f_{\mathbf{DC}}(\mathbb{B}_1,\mathbb{B}_2)}$

$$E_{f_{\mathbf{DC}}(\mathbb{B}_1,\mathbb{B}_2)} = \begin{cases} 0, & f_{\mathbf{DC}}(\mathbb{B}_1,\mathbb{B}_2) \le 0 \\ \frac{2}{1+e^{-f_{\mathbf{DC}}(\mathbb{B}_1,\mathbb{B}_2)}} - 1, & f_{\mathbf{DC}}(\mathbb{B}_1,\mathbb{B}_2) > 0 \end{cases}$$

Given Ball $\mathbb{B}$, $\mathbb{B}_0$ is inside $\mathbb{B}$, and $\mathbb{B}_i$ disconnects from $\mathbb{B}$, in which $1 \le i \le N$. The Loss Function $L_{\mathbb{B}}$ is defined as

$$L_{\mathbb{B}} = E_{f_{\mathbf{P}}(\mathbb{B}_0,\mathbb{B})} + \sum_{i=1}^{N} E_{f_{\mathbf{DC}}(\mathbb{B}_i,\mathbb{B})} \tag{4.57}$$

4.4.2.3 Partial Derivatives for the Back Propagation Method

We use back-propagation method to update radii and lengths of central vectors, while keeping directions of central vectors.

$$\begin{aligned}
\frac{\partial L_{\mathbb{B}}}{\partial l_{\mathbb{B}}} &= \frac{\partial E_{f_{\mathbf{P}}(\mathbb{B}_0,\mathbb{B})}}{\partial l_{\mathbb{B}}} + \sum_{i=1}^{N} \frac{\partial E_{f_{\mathbf{DC}}(\mathbb{B}_i,\mathbb{B})}}{\partial l_{\mathbb{B}}} \\
&= 2(1 - E_{f_{\mathbf{P}(\mathbb{B}_0,\mathbb{B})}}) E_{f_{\mathbf{P}(\mathbb{B}_0,\mathbb{B})}} \frac{\partial f_{\mathbf{P}}(\mathbb{B}_0,\mathbb{B})}{\partial l_{\mathbb{B}}} \\
&\quad + \sum_{i=1}^{N} 2(1 - E_{f_{\mathbf{DC}}(\mathbb{B}_0,\mathbb{B})}) E_{f_{\mathbf{DC}}(\mathbb{B}_0,\mathbb{B})} \frac{\partial f_{\mathbf{DC}}(\mathbb{B}_i,\mathbb{B})}{\partial l_{\mathbb{B}}} \\
&= 2(1 - E_{f_{\mathbf{P}(\mathbb{B}_0,\mathbb{B})}}) E_{f_{\mathbf{P}(\mathbb{B}_0,\mathbb{B})}} \frac{l_{\mathbb{B}} - l_{\mathbb{B}_0} \boldsymbol{\alpha}_{\mathbb{B}_0} \cdot \boldsymbol{\alpha}_{\mathbb{B}}}{\sqrt{l_{\mathbb{B}}^2 + l_{\mathbb{B}_0}^2 - 2 l_{\mathbb{B}} l_{\mathbb{B}_0} \boldsymbol{\alpha}_{\mathbb{B}} \cdot \boldsymbol{\alpha}_{\mathbb{B}_0}}} \\
&\quad + \sum_{i=1}^{N} 2(1 - E_{f_{\mathbf{DC}}(\mathbb{B}_0,\mathbb{B})}) E_{f_{\mathbf{DC}}(\mathbb{B}_0,\mathbb{B})} \frac{l_{\mathbb{B}} - l_{\mathbb{B}_i} \boldsymbol{\alpha}_{\mathbb{B}_i} \cdot \boldsymbol{\alpha}_{\mathbb{B}}}{\sqrt{l_{\mathbb{B}}^2 + l_{\mathbb{B}_i}^2 - 2 l_{\mathbb{B}} l_{\mathbb{B}_i} \boldsymbol{\alpha}_{\mathbb{B}} \cdot \boldsymbol{\alpha}_{\mathbb{B}_i}}}
\end{aligned}$$

$$\begin{aligned}
\frac{\partial L_{\mathbb{B}}}{\partial r_{\mathbb{B}}} &= 2(1 - E_{f_{\mathbf{P}(\mathbb{B}_0,\mathbb{B})}}) E_{f_{\mathbf{P}(\mathbb{B}_0,\mathbb{B})}} \frac{\partial f_{\mathbf{P}}(\mathbb{B}_0,\mathbb{B})}{\partial r_{\mathbb{B}}} \\
&\quad + \sum_{i=1}^{N} 2(1 - E_{f_{\mathbf{DC}}(\mathbb{B}_0,\mathbb{B})}) E_{f_{\mathbf{DC}}(\mathbb{B}_0,\mathbb{B})} \frac{\partial f_{\mathbf{DC}}(\mathbb{B}_i,\mathbb{B})}{\partial r_{\mathbb{B}}} \\
&= (2E_{f_{\mathbf{P}(\mathbb{B}_0,\mathbb{B})}} - 2) E_{f_{\mathbf{P}(\mathbb{B}_0,\mathbb{B})}} + \sum_{i=1}^{N} (2 - 2E_{f_{\mathbf{DC}}(\mathbb{B}_i,\mathbb{B})}) E_{f_{\mathbf{DC}}(\mathbb{B}_i,\mathbb{B})}
\end{aligned}$$

$$\frac{\partial L_{\mathbb{B}}}{\partial l_{\mathbb{B}_0}} = 2(1 - E_{f_{\mathbf{P}(\mathbb{B}_0,\mathbb{B})}})E_{f_{\mathbf{P}(\mathbb{B}_0,\mathbb{B})}}\frac{\partial f_{\mathbf{P}}(\mathbb{B}_0,\mathbb{B})}{\partial l_{\mathbb{B}_0}}$$
$$+\sum_{i=1}^{N} 2(1 - E_{f_{\mathbf{DC}}(\mathbb{B}_0,\mathbb{B})})E_{f_{\mathbf{DC}}(\mathbb{B}_0,\mathbb{B})}\frac{\partial f_{\mathbf{DC}}(\mathbb{B}_i,\mathbb{B})}{\partial l_{\mathbb{B}_0}}$$
$$= (2 - 2E_{f_{\mathbf{P}(\mathbb{B}_0,\mathbb{B})}})E_{f_{\mathbf{P}(\mathbb{B}_0,\mathbb{B})}}\frac{l_{\mathbb{B}_0} - l_{\mathbb{B}}\boldsymbol{\alpha}_{\mathbb{B}_0}\cdot\boldsymbol{\alpha}_{\mathbb{B}}}{\sqrt{l_{\mathbb{B}}^2 + l_{\mathbb{B}_0}^2 - 2l_{\mathbb{B}}l_{\mathbb{B}_0}\boldsymbol{\alpha}_{\mathbb{B}}\cdot\boldsymbol{\alpha}_{\mathbb{B}_0}}}$$
$$\frac{\partial L_{\mathbb{B}}}{\partial r_{\mathbb{B}_0}} = 2(1 - E_{f_{\mathbf{P}(\mathbb{B}_0,\mathbb{B})}})E_{f_{\mathbf{P}(\mathbb{B}_0,\mathbb{B})}}$$

$$\frac{\partial L_{\mathbb{B}}}{\partial l_{\mathbb{B}_i}} = \sum_{i=1}^{N} 2(1 - E_{f_{\mathbf{DC}}(\mathbb{B}_i,\mathbb{B})})E_{f_{\mathbf{DC}}(\mathbb{B}_i,\mathbb{B})}\frac{\partial f_{\mathbf{DC}}(\mathbb{B}_i,\mathbb{B})}{\partial l_{\mathbb{B}_i}}$$
$$= (2E_{f_{\mathbf{DC}}(\mathbb{B}_i,\mathbb{B})} - 2)E_{f_{\mathbf{DC}}(\mathbb{B}_i,\mathbb{B})}\frac{l_{\mathbb{B}_i} - l_{\mathbb{B}}\boldsymbol{\alpha}_{\mathbb{B}_i}\cdot\boldsymbol{\alpha}_{\mathbb{B}}}{\sqrt{l_{\mathbb{B}}^2 + l_{\mathbb{B}_i}^2 - 2l_{\mathbb{B}}l_{\mathbb{B}_i}\boldsymbol{\alpha}_{\mathbb{B}}\cdot\boldsymbol{\alpha}_{\mathbb{B}_i}}}$$
$$\frac{\partial L_{\mathbb{B}}}{\partial r_{\mathbb{B}_i}} = \sum_{i=1}^{N} 2(1 - E_{f_{\mathbf{DC}}(\mathbb{B}_i,\mathbb{B})})E_{f_{\mathbf{DC}}(\mathbb{B}_i,\mathbb{B})}\frac{\partial f_{\mathbf{DC}}(\mathbb{B}_i,\mathbb{B})}{\partial r_{\mathbb{B}_i}}$$
$$= (2E_{f_{\mathbf{DC}}(\mathbb{B}_i,\mathbb{B})} - 2)E_{f_{\mathbf{DC}}(\mathbb{B}_i,\mathbb{B})}$$

4.4.3 Experiment Results and Analysis

We implemented the back-propagation approach in Python. The source code is public accessible at github https://github.com/gnodisnait/bp94nball. The back-propagation approach cannot guarantee to achieve our target configuration with zero-energy cost. Sometimes it can achieve zero-energy cost, as shown in Fig. 4.9, more often than not, ending up with an unintended configuration, as illustrated in Fig. 4.10. The reason lies in the fact that the back-propagation approach is to maximize a harmony (*harmony maxima*) between input-output data (a typical function of System 1) (Smolensky 1988; Freeman 1988), and inevitably breaks already improved relations. In contrast, our target is to precisely encode symbolic structures, which is a task of *veridical representation* (Smolensky 1988).

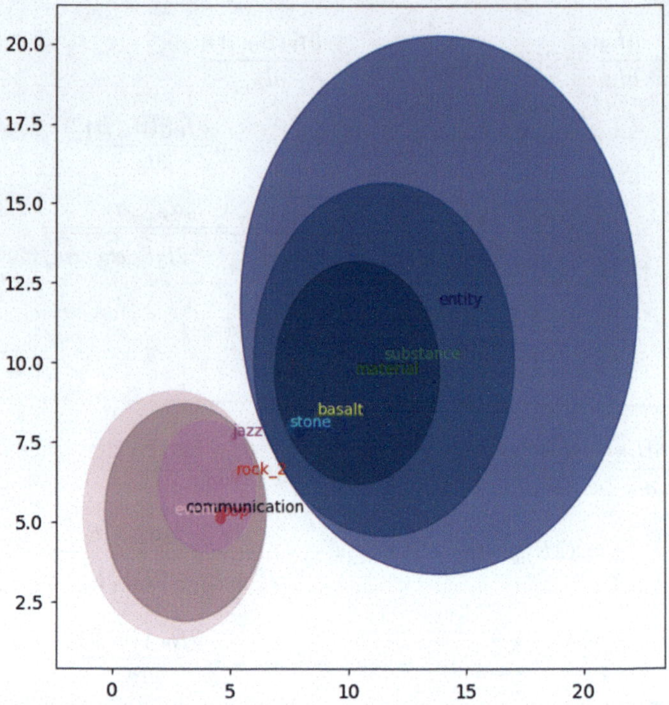

Fig. 4.9 The back-propagation approach sometimes achieves zero-energy cost

4.5 Summary

In this chapter, we describe the task of spatializing symbolic tree structures onto vector embeddings, list the criteria and the challenge. We review the fundamental algorithm in Deep Learning–the back-propagation method, and demonstrate its magic power in completing route instructions. Through a concrete example of spatializing two symbolic tree structures, we conclude that this method is not the right approach to spatializing symbolic structures.

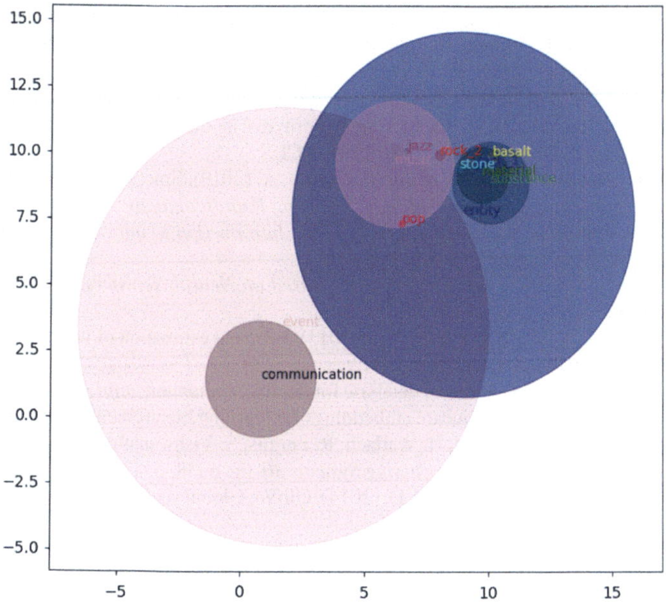

Fig. 4.10 The back-propagation approach fails to achieve zero-energy cost

References

Adamson, M. J., & Damper, R. I. (1999). B-RAAM: A connectionist model which develops holistic internal representations of symbolic structures. *Connection Science, 11*(1), 41–71.

Athiwaratkun, B., & Wilson, A. (2017). Multimodal word distributions. In *ACL'17*, pp. 1645–1656.

Bengio, Y., Ducharme, R., Vincent, P., & Janvin, C. (2003). A neural probabilistic language model. *Journal of Machine Learning Research, 3*, 1137–1155.

Collobert, R., Weston, J., Bottou, L., Karlen, M., Kavukcuoglu, K., & Kuksa, P. (2011). Natural language processing (almost) from scratch. *Journal of Machine Learning Research, 12*, 2493–2537.

Erk, K. (2009). Supporting inferences in semantic space: Representing words as regions. In *IWCS-8'09*, pp. 104–115, Stroudsburg, PA, USA. Association for Computational Linguistics.

Faruqui, M., Dodge, J., Jauhar, S. K., Dyer, C., Hovy, E., & Smith, N. A. (2015). Retrofitting word vectors to semantic lexicons. In *Proceedings of the 2015 Conference of the North American Chapter of the Association for Computational Linguistics: Human Language Technologies*, pp. 1606–1615. ACL.

Freeman, W. J. (1988). Dynamic systems and the "subsymbolic level". *Behavioral and Brain Sciences, 1*, 33–34.

Fu, R., Guo, J., Qin, B., Che, W., Wang, H., & Liu, T. (2015). Learning semantic hierarchies: A continuous vector space approach. *Transactions Audio, Speech and Language Proceedings, 23*(3):461–471.

Han, X., Liu, Z., & Sun, M. (2016). Joint representation learning of text and knowledge for knowledge graph completion. CoRR arXiv:abs/1611.04125.

Kelley, J. K. (1955). *General topology*. New York: Springer.

Kruszewski, G., Paperno, D., & Baroni, M. (2015). Deriving boolean structures from distributional vectors. *Transactions of the Association of Computational Linguistics, 3*, 375–388.

LeCun, Y., Chopra, S., Hadsell, R., Ranzato, M., & Huang, F. (2006). A tutorial on energy-based learning. In G. Bakir, T. Hofman, B. Schölkopf, A. Smola, & B. Taskar (Eds.), *Predicting Structured Data*. MIT Press.
LeCun, Y., Bengio, Y., & Hinton, G. E. (2015). Deep learning. *Nature*, *521*(7553), 436–444.
Lenci, A., & Benotto, G. (2012). Identifying hypernyms in distributional semantic spaces. In *SemEval '12*, pp. 75–79, Stroudsburg, PA, USA. ACL.
Li, X., Vilnis, L., Zhang, D., Boratko, M., & McCallum, A. (2019). Smoothing the geometry of box embeddings. In *International Conference on Learning Representations (ICLR)*.
Manning, C. D., Raghavan, P., & Schütze, H. (2008). *Introduction to information retrieval*. New York, NY, USA: Cambridge University Press.
Mikolov, T. (2012). *Statistical Language Models Based on Neural Networks*. Ph.D. thesis, Brno University of Technology, Brno, CZ.
Mikolov, T., Chen, K., Corrado, G., & Dean, J. (2013). Efficient estimation of word representations in vector space. CoRR arXiv:abs/1301.3781.
Miller, G. A. (1995). Wordnet: A lexical database for english. *Communication ACM*, *38*(11), 39–41.
Nickel, M., & Kiela, D. (2017). Poincaré embeddings for learning hierarchical representations. In I. Guyon, U. V. Luxburg, S. Bengio, H. Wallach, R. Fergus, S. Vishwanathan, & R. Garnett (Eds.) *Advances in Neural Information Processing Systems 30*, pp. 6338–6347. Curran Associates, Inc.
Pennington, J., Socher, R., & Manning, C. D. (2014). GloVe: Global vectors for word representation. In *EMNLP'14*, pp. 1532–1543.
Pollack, J. B. (1990). Recursive distributed representations. *Artificial Intelligence*, *46*(1–2), 77–105.
Rumelhart, D. E., McClelland, J. L., & PDP Research Group, C. (Eds.) (1986). *Parallel distributed processing: Explorations in the microstructure of cognition, Vol. 1: Foundations*. MIT Press, Cambridge, MA, USA.
Sebastiani, F. (2002). Machine learning in automated text categorization. *ACM Computing Surveys*, *34*(1), 1–47.
Smolensky, P. (1988). On the proper treatment of connectionism. *Behavioral and Brain Sciences*, *1*, 1–23.
Socher, R., Chen, D., Manning, C. D., & Ng, A. (2013). Reasoning with neural tensor networks for knowledge base completion. In C. J. C. Burges, L. Bottou, M. Welling, Z. Ghahramani, & K. Q. Weinberger (Eds.) *Advances in Neural Information Processing Systems 26*, pp. 926–934. Curran Associates, Inc.
Speer, R., Chin, J., & Havasi, C. (2017). Conceptnet 5.5: An open multilingual graph of general knowledge. In *Proceedings of the Thirty-First AAAI Conference on Artificial Intelligence, February 4-9, 2017, San Francisco, California, USA.*, pp. 4444–4451.
Tellex, S., Katz, B., Lin, J., Fernandes, A., & Marton, G. (2003). Quantitative evaluation of passage retrieval algorithms for question answering. In *Proceedings of the 26th Annual International ACM SIGIR Conference on Research and Development in Information Retrieval*, SIGIR '03, pp. 41–47, New York, NY, USA. ACM.
Tversky, B. (2019). *Mind in motion*. New York, USA: Basic Books.
Xiao, H., Huang, M., & Zhu, X. (2016). From one point to a manifold: Knowledge graph embedding for precise link prediction. *IJCAI*, 1315–1321.

Chapter 5
Design Principles of Geometric Connectionist Machines

In this chapter, we introduce a novel geometric method to precisely spatialize symbolic tree structures onto vector embeddings.

5.1 Principle of Family Action (PFA)

The lesson that we learned in the last chapter is that we shall prevent already improved relations from deteriorating in the later updating process. A simple remedy will be the Principle of Family Action (PFA): *If an updating operation is applied for a ball, the same operation shall be applied for all its descendant balls.*

5.2 Principle of Depth First (PDF)

The problem introduced by PFA is the computational complexity. Each time a ball is updated, all its descendant balls will be updated. To reduce the number of updating times for each ball, we need the Principle of Depth First (PDF): *The updating of child balls is performed before the updating of their parent ball.* This follows the depth first recursive process. There will be 4 possible sequences to traverse the tree structure in Fig. 5.1 as follows.

1. apple, google, company, religion, social_group
2. google, apple, company, religion, social_group
3. religion, apple, google, company, social_group
4. religion, google, apple, company, social_group

T. Dong, *A Geometric Approach to the Unification of Symbolic Structures and Neural Networks*, Studies in Computational Intelligence 910,
https://doi.org/10.1007/978-3-030-56275-5_5

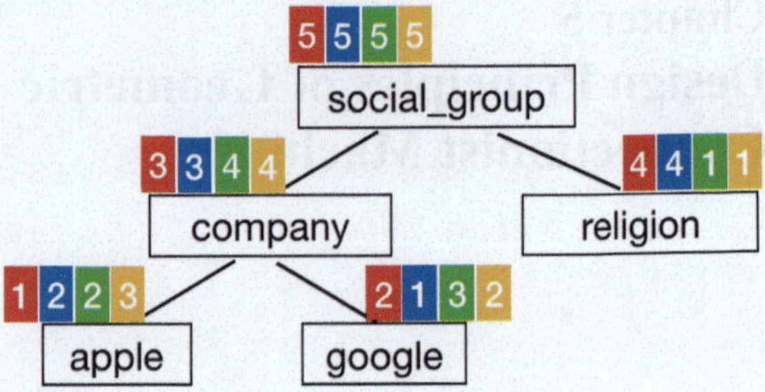

Fig. 5.1 Each color marks one possible sequence to traverse the tree

5.3 Principle of Large Sibling Family First (PLSFF)

The criteria of Zero-Energy Cost requires that all sibling balls are mutually disconnected. Suppose that company node is constructed after religion node (either the green or the yellow sequence in Fig. 5.1), and that company ball connects with religion ball. We need to move away company ball to make it disconnect from religion ball. Following the principle of PFA, we shall apply the same movement for apple ball and google ball. The number of total geometric operations will be larger than that of sequences in which company ball is constructed before that religion ball (either the red or the blue sequence in Fig. 5.1). This triggers out the Principle of Large Sibling Family First (PLSFF): *Among all sibling balls, the ball having the larger number of descendant balls shall be constructed before sibling balls having less number of descendant balls.*

5.4 Principle of Homothetic Transformation First (PHTF)

Now, let us follow the above principles and go through the geometric construction process of the tree in Fig. 5.1. Without the loss of generality, we choose the red sequence, and firstly initialize apple ball, then initialize google ball. Suppose that google ball connects with apple ball, as shown in Fig. 5.2a, we need to figure out a geometric transformation for google ball. As we want to preserve the pre-trained vector, we would move away google ball in the direction of its central point vector, as shown in Fig. 5.2b. Following the Principle of Family Action, all balls contained by google ball shall be applied with the same transformation. To preserve the inclusion relations from deteriorating, we shall enlarge their radii with the same ratio as the length change of their central vectors. This is the homothetic transformation defined as follows.

Definition Given Ball $\mathbb{B}(\mathbf{O}, r) = \mathbb{B}((\boldsymbol{\alpha}, l), r)$, the Homothetic operation on $\mathbb{B}$ with the ratio $k(k > 0)$, written as $\mathsf{H}(\mathbb{B}, k)$, transforms $\mathbb{B}((\boldsymbol{\alpha}, l), r)$ into $\mathbb{B}((\boldsymbol{\alpha}, kl), kr)$. □

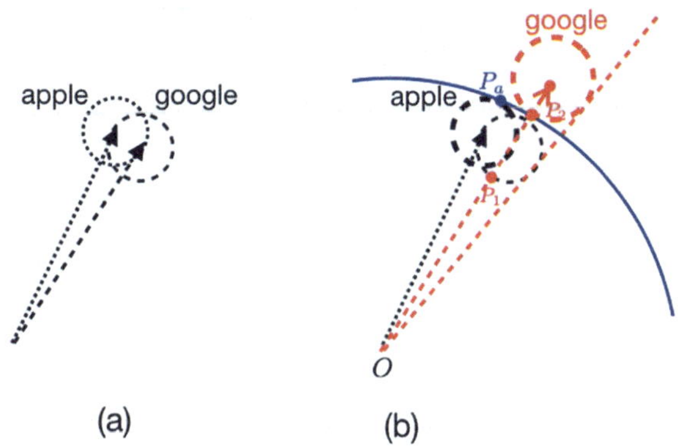

Fig. 5.2 **a** Google ball connects with apple ball; **b** A homothetic transformation is applied for google ball

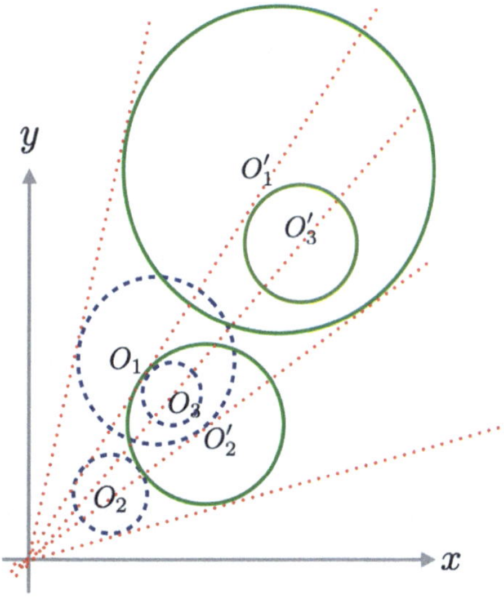

Fig. 5.3 Ball O_1 disconnects from ball O_2, and contains ball O_3. After homothetic transformation, both relations are kept: ball O_1' disconnects from ball O_2', and contains ball O_3'

Homothetic transformation preserves the relation of being inside and the relation of being disconnected, as illustrated in Fig. 5.3. We briefly prove these features as follows.

Theorem *Given Ball* $\mathbb{B}_1(\mathbf{O}_1, r_1) = \mathbb{B}_1((\boldsymbol{\alpha}_1, l_1), r_1)$ *and Ball* $\mathbb{B}_2(\mathbf{O}_2, r_2) = \mathbb{B}_1((\boldsymbol{\alpha}_2, l_2), r_2)$. *If* $\mathbf{P}(\mathbb{B}_1, \mathbb{B}_2)$, *then* $\mathbf{P}(\mathsf{H}(\mathbb{B}_1, k), \mathsf{H}(\mathbb{B}_2, k))$, *in which* $k > 0$.

□

Proof With $\mathbf{P}(\mathbb{B}_1, \mathbb{B}_2)$, we have $f_{\mathbf{P}}(\mathbb{B}_1, \mathbb{B}_2) \triangleq \|l_2 \cdot \boldsymbol{\alpha_2} - l_1 \cdot \boldsymbol{\alpha_1}\| + r_1 - r_2 \leq 0$.

$$
\begin{aligned}
f_{\mathbf{P}}(\mathsf{H}(\mathbb{B}_1, k), \mathsf{H}(\mathbb{B}_2, k)) &\triangleq \|(kl_2) \cdot \boldsymbol{\alpha_2} | (kl_1) \cdot \boldsymbol{\alpha_1}\| + kr_1 - kr_2 \\
&= k(\|l_2 \cdot \boldsymbol{\alpha_2} - l_1 \cdot \boldsymbol{\alpha_1}\| + r_1 - r_2) \\
&= kf_{\mathbf{P}}(\mathbb{B}_1, \mathbb{B}_2) \\
&\leq 0
\end{aligned}
$$

Theorem *Given Ball* $\mathbb{B}_1(\mathbf{O}_1, r_1) = \mathbb{B}_1((\boldsymbol{\alpha}_1, l_1), r_1)$ *and Ball* $\mathbb{B}_2(\mathbf{O}_2, r_2) = \mathbb{B}_2((\boldsymbol{\alpha}_2, l_2), r_2)$. *If* $\mathbf{DC}(\mathbb{B}_1, \mathbb{B}_2)$, *then* $\mathbf{DC}(\mathsf{H}(\mathbb{B}_1, k), \mathsf{H}(\mathbb{B}_2, k))$, *in which* $k > 0$. □

Proof As $\mathbf{DC}(\mathbb{B}_1, \mathbb{B}_2)$, we have $f_{\mathbf{DC}}(\mathbb{B}_1, \mathbb{B}_2) \triangleq r_1 + r_2 - \|l_2 \cdot \boldsymbol{\alpha_2} - l_1 \cdot \boldsymbol{\alpha_1}\| \leq 0$.

$$\begin{aligned} f_{\mathbf{DC}}(\mathsf{H}(\mathbb{B}_1, k), \mathsf{H}(\mathbb{B}_2, k)) &\triangleq kr_1 + kr_2 - \|kl_2 \cdot \boldsymbol{\alpha_2} - kl_1 \cdot \boldsymbol{\alpha_1}\| \\ &= k(r_1 + r_2 - \|l_2 \cdot \boldsymbol{\alpha_2} - l_1 \cdot \boldsymbol{\alpha_1}\|) \\ &= k f_{\mathbf{DC}}(\mathbb{B}_1, \mathbb{B}_2) \\ &\leq 0 \end{aligned}$$

With this nice feature of the homothetic transformation, we propose the Principle of Homothetic Transformation (PHTF): *The homothetic transformation has the priority than other geometric transformations.* We can use homothetic transformation to separate balls, if they do not contain the origin point of the space.

Theorem *Given Ball* $\mathbb{B}(\mathbf{O}, r) = \mathbb{B}((\boldsymbol{\alpha}, l), r)$ *containing the origin point* ($l < r$), *we have* $\mathbf{P}(\mathbb{B}, \mathsf{H}(\mathbb{B}, k))$, *in which* $k > 1$. □

Proof

$$\begin{aligned} f_{\mathbf{P}}(\mathbb{B}, \mathsf{H}(\mathbb{B}, k)) &\triangleq \|kl\boldsymbol{\alpha} - l\boldsymbol{\alpha}\| + kr - r \\ &= (k-1)(l-r) \\ &< 0 \end{aligned}$$

therefore, $\mathbf{P}(\mathbb{B}, \mathsf{H}(\mathbb{B}, k))$ □

Theorem *Given Ball* $\mathbb{B}(\mathbf{O}, r) = \mathbb{B}((\boldsymbol{\alpha}, l), r)$ *containing the origin point* ($l < r$), *we apply homothetic transformation for* $\mathbb{B}(\mathbf{O}, r)$, $\mathsf{H}(\mathbb{B}, k)$, *in which* $k > 1$. *For any Ball* $\mathbb{B}(\mathbf{O}', r')$, *if* $\mathbb{B}(\mathbf{O}', r')$ *connects with* $\mathbb{B}(\mathbf{O}, r)$, $\mathbb{B}(\mathbf{O}', r')$ *also connects with* $\mathsf{H}(\mathbb{B}, k)$ □

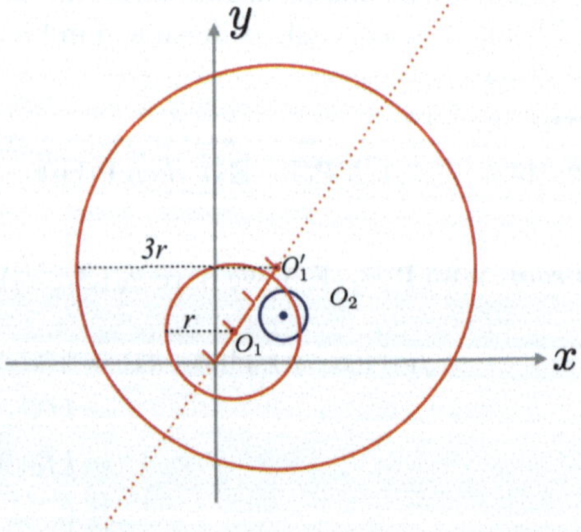

Fig. 5.4 Ball O_1 contains the origin point O, and overlaps with Ball O_2. After we apply the homothetic transformation for Ball O_1 with $k = 3$, it still overlaps with Ball O_2

Figure 5.4 illustrates the case that the homothetic transformation fails to separate ball O_1 from Ball O_2, as Ball O_1 contains the original point of the space ($l \leq r$). In this case, we have to shift the ball away from the origin point, so that $l > r$.

5.5 Principle of Shifting and Rotation Transformation (PSRT)

Definition Given Ball $\mathbb{B}(\mathbf{O}, r) = \mathbb{B}((\boldsymbol{\alpha}, l), r)$, the Shifting operation on $\mathbb{B}$ with vector $\mathbf{s}$, written as $\mathsf{S}(\mathbb{B}, \mathbf{s})$, transforms $\mathbb{B}((\boldsymbol{\alpha}, l), r)$ into $\mathbb{B}(l\boldsymbol{\alpha} + \mathbf{s}, r)$. □

To keep pre-trained vectors from being changed, we propose to shift in the direction of the central point vector. However, following the Principle of Family Action, the central point vectors of child balls will be changed, as shown in Fig. 5.5.

The last geometric transformation is the Rotation transformation, which keeps the length of the radius r and the length of the central vector. The Rotation transformation rotates the orientation of the central vector with angle β in the subspace spanned by the i-th and the j-th basis. As we would like to preserve the pre-trained vectors, the subspace for rotation shall be selected inside the extended space.

Definition Given Ball $\mathbb{B}(\mathbf{O}, r) = \mathbb{B}((\boldsymbol{\alpha}, l), r)$, the Rotation operation on $\mathbb{B}$ with unit vector $\boldsymbol{\beta}$ in the subspace spanned by the i-th and the j-th basis, written as $\mathsf{R}(\mathbb{B}, \boldsymbol{\beta}, i, j)$, transforms $\mathbb{B}((\boldsymbol{\alpha}, l), r)$ into $\mathbb{B}'((\boldsymbol{\alpha}', l), r)$. Let $\boldsymbol{\alpha} = [e_1, \ldots, e_n]$ and $\boldsymbol{\alpha}' = [e'_1, \ldots, e'_n]$, $e_k = e'_k$ for all $k \notin \{i, j\}$, $e'_i = e_i \cos\beta + e_j \sin\beta$, $e'_j = -e_i \sin \beta + e_j \cos\beta$. □

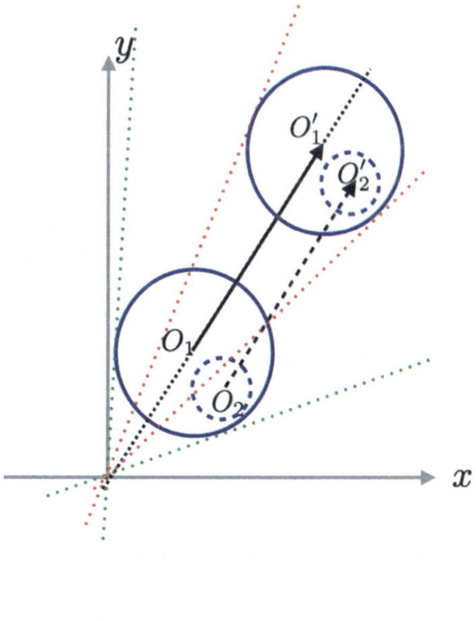

Fig. 5.5 A shift transformation on Ball O_1 will change the central point vectors of child balls, e.g. Ball O_2

The Rotation transformation will not help to push a ball away from the origin point, and has the lower priority than the Shifting transformation. The Principle of Shifting and Rotation Transformation is as follows: *Shifting transformation has higher priority than Rotation transformation.*

5.6 Principle of Increasing Dimensions (POID)

Erk (2009) extended word embeddings into regions to predict hypernym relations in the co-occurrence space. She did experiments with 120 monosemous verbs (with 430 hypernyms) and found that words do not always fall inside their hypernym regions. Her work suggested that the embedding space of hypernym relations shall not be exactly the same as the embedding space of co-occurrence relations. Similar work has been carried out by Fu et al. (2015), who used embedding approach to automatically predict subordinate relations among Chinese words. Their experiments achieved F-score of 73.74%, and if manually labelled subordinate relations are used, the F-score will increase to 80.29%. Erk (2009) and Fu et al. (2015) have the convergent observation that hyponym-hypernym word pairs do not always appear in the same context. This is quite understandable. As symbolic relations, hyponym-hypernym word pairs are manually identified by linguists who refer to a variety of resources. In contrast, connectionists or statisticians use a large amount of text resources. Symbolic relations may convey information beyond the reach of the corpus used by connectionists or statisticians. In term of representation, the space of symbolic relations may have different dimensions from the space learned from corpus. This follows that the space structured by hyponym relations should have new dimensions. The Principle of Increasing Dimensions is stated as follows: *The dimensions of the vector space shall be increased, in order to precisely spatialize symbolic relations among entities onto the vector space structured by connectionist networks.* This can be illustrated in the simple example of extracting hyponym words from corpus as follows. If we know that Hamburg is a harbor city, what are other harbor cities? If we do not add dimensions to word-embeddings, what we can do is to draw a circle with harbor city as the central point. This circle should contain the vector of Hamburg. However, this circle also contains vectors of Berlin, capital, and city, as illustrated in Fig. 5.6a. This may help us to understand the reason that the recall and the precision of hyponym extraction are not as high as we expected. In contrast, if we add a new dimension, and turn the pre-trained word vectors into two dimensional circles, it would be possible to precisely encode hyponym relations in terms of inclusion relations among circles, as illustrated in Fig. 5.6b.

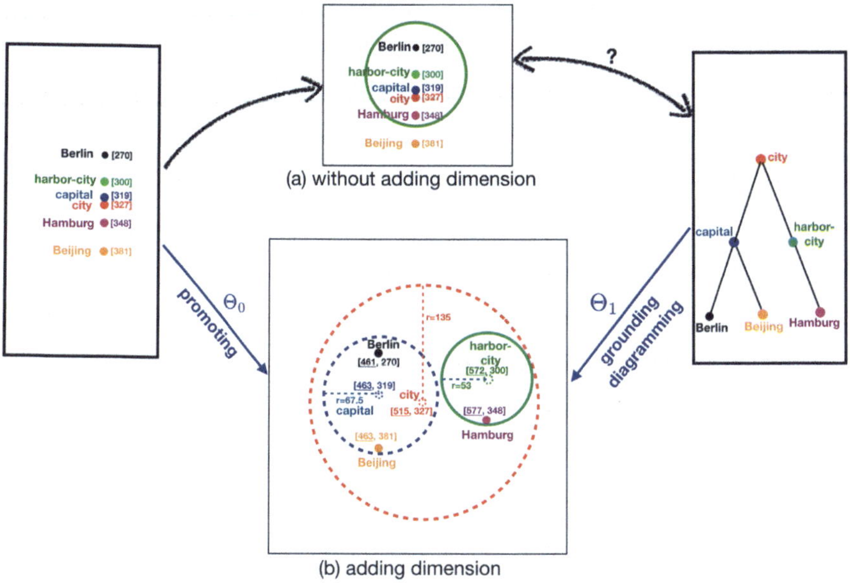

Fig. 5.6 **a** Without adding a new dimension, hyponyms of harbor-city include Hamburg, as well as Berlin, capital, and city; **b** by adding a new dimension, the hyponyms of harbor-city only contain Hamburg

5.7 Geometric Approach to Spatializing a Tree Structure

The first operation of symbol spatialization is the promoting processing Θ_0, as illustrated in Fig. 3.9 in Sect. 3.5. Following the Principle of Increasing Dimensions, this operation will promote a vector onto an $\mathcal{N}$-Ball in a higher dimensional space.

Definition Let $\mathfrak{T}$ be a tree structure, t be a node in $\mathfrak{T}$, $\mathbf{v}_t$ and $\mathbf{w}_t$ be pre-trained and extended vectors of t. Given two positive real numbers d_0 and $r_0 = 0^+$, the promoting operator Θ_0 initializes an $\mathcal{N}$-Ball $\mathbb{B}((c_t, d_0), r_0)$ of t, in which the direction of c_t is constructed by concatenating $\mathbf{v}_t$ and $\mathbf{w}_t$, $\mathbf{v}_t^\frown \mathbf{w}_t$, as follows.

$$c_t = d_0 \cdot \frac{\mathbf{v}_t^\frown \mathbf{w}_t}{\|\mathbf{v}_t^\frown \mathbf{w}_t\|} \tag{5.1}$$

$$\Theta_0(t, \mathbf{v}_t, \mathbf{w}_t) = \{p | \, \|p - c_t\| < r_t\} \tag{5.2}$$

□

We use depth-first algorithm to compute the inclusion relations between child-parent balls and the disconnectedness relations among sibling balls. The spatializing procedure Θ(Fig. 3.9) can be implemented by the `spatializing_tree` procedure, as listed in Algorithm 1. The procedure can be described as follows: it goes to the leaf node first, and initialize a ball with default radius and the center point;

then recursively moves to all its sibling branches and does the same; after processed all the sibling branches, it shall guarantee that all sibling balls disconnect from each other, then goes to upper-level to construct the parent ball.

Algorithm 1: Depth first algorithm to spatialize tree Tree (with root as its root node), `spatializing_tree`(root, Tree, V, W, Balls)

```
input : a Tree structure pointed by root
input : Dictionary of pre-trained node embeddings V
input : Dictionary of extended node embeddings W
output: Dictionary of created 𝒩-Ball embeddings Balls

children ⟵ get_children_of(root, Tree)
if len(children) > 0 then
    foreach child ∈ children do
        // depth first
        Balls = spatializing_tree(child, Tree, V, W, Balls)
    end
    if len(children) > 1 then
        // updating siblings to be disconnected from each other
        Balls = update_to_be_disconnected(children, Balls)
    end
    // create parent ball for all children
    Balls [root ] = create_parent_ball(root, children, Balls)
else
    // initializing a vector onto an 𝒩-Ball
    Balls [root ] = Θ0(root, V [root ], W [root ])
end
```

5.7.1 Separating Sibling Balls Apart

To guarantee all sibling balls to be mutually separated, we will pair-wisely check sibling balls, and separate them, if they are connected. With the Principle of Large Sibling Family First (PLSFF), we decreasingly sort them by counting the total family number of each ball. Formally, let $\mathbb{B}_1, \mathbb{B}_2, \ldots, \mathbb{B}_N$ be siblings, and the number of $\mathcal{N}$-Balls contained by $\mathbb{B}_i$ is no less than that of $\mathcal{N}$-Balls contained by $\mathbb{B}_j$, in which $i < j$.

The Principle of Family Action (PFA) requires that the same geometric transformations shall be applied for all $\mathcal{N}$-Balls contained by $\mathbb{B}_k$, after $\mathbb{B}_k$ is updated. Suppose $i < j < k$, and $\mathbb{B}_k$ is updated during both the $i \sim j$ loop and the $j \sim k$ loop, all $\mathcal{N}$-Balls contained by $\mathbb{B}_k$ will be updated in both loops. An optimized algorithm is as follows: we firstly make all $\mathbb{B}_i$s being disconnected from each other, and do not update their child balls. After that, apply geometric transformation for all the descendant balls of $\mathbb{B}_i$, according to $\mathbb{B}_i$'s change, as described in Algorithm 3.

Algorithm 2: Separating Siblings to be disconnected from each other `update_to_be_disconnected`([$\mathbb{B}_1, \mathbb{B}_2, \ldots, \mathbb{B}_N$], Balls) (Version 1)

---

input : The list of sibling Balls [$\mathbb{B}_1, \mathbb{B}_2, \ldots, \mathbb{B}_N$]
input : Current Dictionary of created $\mathcal{N}$-Ball embeddings Balls
output: Updated Dictionary of created $\mathcal{N}$-Ball embeddings Balls

```
for i = 1 ... N − 1 do
    for j = i + 1 ... N do
        if B_i connects with B_j then
            Apply geometric transformation for B_j;
            Apply the same geometric transformation for all N-Balls contained by B_j;
        end
    end
end
```

Algorithm 3: Separating Siblings to be disconnected from each other `update_to_be_disconnected`([$\mathbb{B}_1, \mathbb{B}_2, \ldots, \mathbb{B}_N$], Balls) (Version 2)

---

input : The list of sibling Balls [$\mathbb{B}_1, \mathbb{B}_2, \ldots, \mathbb{B}_N$]
input : Current Dictionary of created $\mathcal{N}$-Ball embeddings Balls
output: Updated Dictionary of created $\mathcal{N}$-Ball embeddings Balls

```
copy B_i((α_i, l_i), r_i) to B_i((α'_i, l'_i), r'_i) (1 < i < N)
for i = 1 ... N − 1 do
    for j = i + 1 ... N do
        if B_i connects with B_j then
            Apply geometric transformation for B_j;
        end
    end
end
for i = 2 ... N do
    if l'_i ≠ l_i then
        Apply Homothetic Transformation for each N-Ball contained by B_i with ratio l_i / l'_i;
    end
end
```

5.7.2 The Construction of the Parent Ball

Following the Principle of Depth First (PDF), we shall construct parent ball, after its child balls are being constructed. Let $\mathbb{B}_1((\alpha_1, l_1), r_1), \ldots \mathbb{B}_n((\alpha_n, l_n), r_n)$ be n constructed child balls, and $\mathbb{B}_p((\alpha_p, l_p), r_p)$ be the initialized parent ball. The construction of their parent ball is separated in two steps:

(1) for each child ball $\mathbb{B}_i((\alpha_i, l_i), r_i)$, we construct its parent ball $\mathbb{B}_{p_i}((\alpha_p, l_{p_i}), r_{p_i})$ that tangentially contains $\mathbb{B}_i$, as illustrated in Fig. 5.7b, c. Geometric relations in Fig. 5.7b are listed as follow.

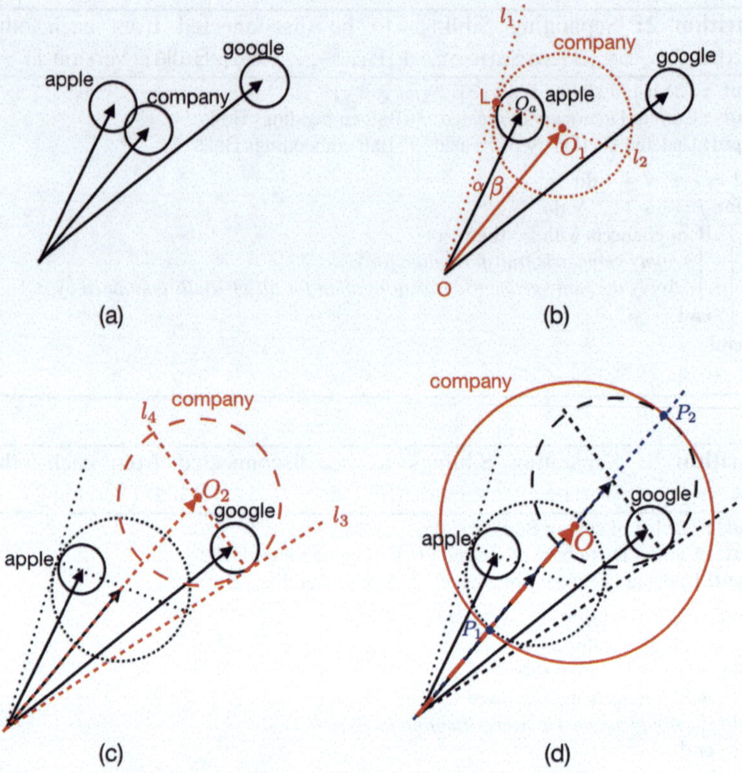

Fig. 5.7 Construct company ball that contains apple ball and google ball

$$|OL| = |\mathbf{O}_a| \cos\alpha \tag{5.3}$$

$$\cos\beta = \frac{\mathbf{O}_a \cdot \mathbf{O}_1}{|\mathbf{O}_a| * |\mathbf{O}_1|} \tag{5.4}$$

$$|OO_1| = \frac{|OL|}{\cos(\alpha+\beta)} \tag{5.5}$$

$$|LO_1| = |OO_1| \sin(\alpha+\beta) \tag{5.6}$$

(2) the final parent ball $\mathbb{B}_p((\alpha_p, l_p), r_p)$ is the minimal cover of all $\mathbb{B}_{p_i}((\alpha_p, l_{p_i}), r_{p_i})$ $(1 < i < n)$. Note all the central points of these $\mathbb{B}_{p_i}$ are in the same orientation pointed by α_p, as illustrated in Fig. 5.7d. We compute l_p and r_p as follow.

$$L_{max} = \max\{l_{p_i} + r_{p_i}\}, i = 1, \ldots, n \tag{5.7}$$

$$L_{min} = \min\{l_{p_i} - r_{p_i}\}, i = 1, \ldots, n \tag{5.8}$$

$$l_p = \frac{L_{\max} + L_{\min}}{2} \tag{5.9}$$

$$r_p = \frac{L_{\max} - L_{\min}}{2} \tag{5.10}$$

5.8 Summary

We present a geometric method which spatializes symbolic tree structures onto vector embeddings of tree nodes with zero energy loss. Each ball represents a symbol satisfying three conditions as follow: (1) the vector representation of the symbol is part of the central vector of the ball; (2) child-parent relations in tree structures are precisely represented by the inclusion relations between the child ball and the parent ball; (3) sibling balls are disconnected from each other. However, the constructed parent ball is not the minimal ball that contains all its descendant balls.

References

Erk, K. (2009). Supporting inferences in semantic space: Representing words as regions. In *IWCS-8'09* (pp. 104–115). Stroudsburg, PA, USA: Association for Computational Linguistics.

Fu, R., Guo, J., Qin, B., Che, W., Wang, H., & Liu, T. (2015). Learning semantic hierarchies: A continuous vector space approach. *IEEE Transactions on Audio, Speech, and Language Processing, 23*(3), 461–471.

Chapter 6
A Geometric Connectionist Machine for Word-Senses

In this chapter, we follow the design principles to precisely spatialize tree structured hypernym relations among word-senses onto word embeddings. Our target is to promote each word-embedding into a ball in a higher dimension space ($\mathcal{N}$-Ball) such that the configuration of these balls precisely captures tree-structured hypernym relations. Each $\mathcal{N}$-Ball represents a word-sense. One $\mathcal{N}$-Ball is contained by another $\mathcal{N}$-Ball, if and only if the word-sense represented by the first $\mathcal{N}$-Ball is a hyponym of the word-sense represented by the second $\mathcal{N}$-Ball.

6.1 Tree Structured Hypernym Relations of Word-Senses

A word can have a number of word-senses, for example, apple can refer to a fruit, a tree, a computer, or a company. WordNet (Miller 1995) is a large knowledge-graph of word-senses. The three word-senses of flower in WordNet 3.0 are as follows: flower.n.01 referring to a plant cultivated for its blooms or blossoms, flower.n.02 referring to a reproductive organ of angiosperm plants, especially one having showy or colorful parts, flower.n.03 referring to the period of greatest prosperity or productivity. A tree structure of word-senses is illustrated in Fig. 6.1. Our aim is to transform it into nested balls, as illustrated in Fig. 6.2.

In tree structures, each node has a unique path from the root. For example, flower.n.01 has a unique path from the root entity.n.01: [entity.n.01, object.n.01, whole.n.02, organism.n.01, plant.n.02, angiosperm.n.01, flower.n.01]. Each node in the path is located at different layers in the tree. If we alphabetically order sibling word-senses, we can assign a unique number to each word-sense within a layer, as illustrated in Fig. 6.2. A path from the root to a word-sense can be rewritten by replacing word-senses with these numbers. This path uniquely identifies the loca-

T. Dong, *A Geometric Approach to the Unification of Symbolic Structures and Neural Networks*, Studies in Computational Intelligence 910,
https://doi.org/10.1007/978-3-030-56275-5_6

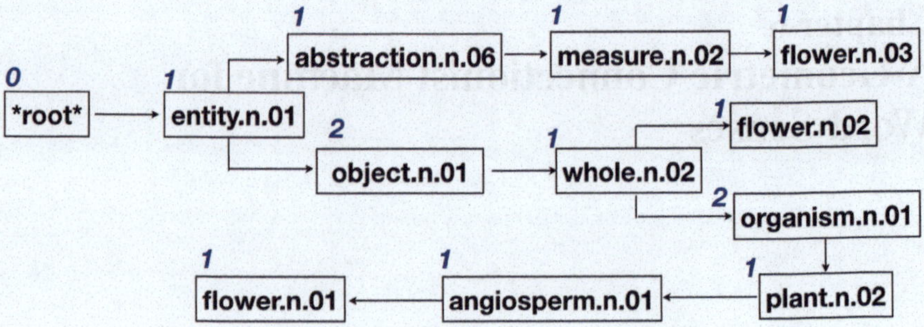

Fig. 6.1 A tree structure in WordNet 3.0 for three word-senses of flower

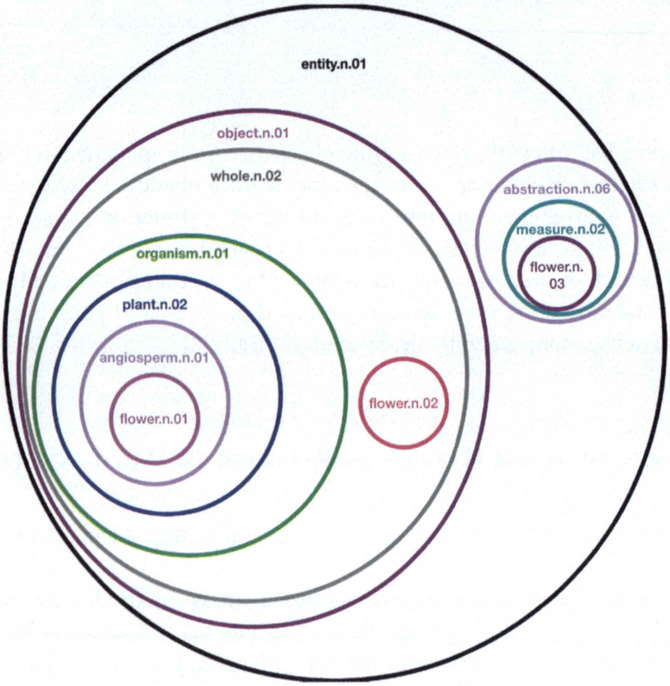

Fig. 6.2 The diagram of $\mathcal{N}$-Ball embeddings of the tree structure in Fig. 6.1

tion of a word-sense in the tree. The direct hypernym of word-sense w_i is its parent node w_{ip} in the tree, which can be uniquely represented by the path of w_{ip}. We call this path *Parent Location Vector (PLV)* of w_i. For example, the *PLV* of flower.n.01 is [1,2,1,2,1,1]; the *PLV* of flower.n.03 is [1,1,1].

6.2 A Geometric Connectionist Machine to Spatialize Unlabeled Tree Structures

We use w_i to represent the ith word-sense of word w, and use $\mathbf{w}$ and $\mathbb{B}_{w_i}$ to represent the pre-trained word-embedding of w and the $\mathscr{N}$-Ball embedding of w_i. $\mathbb{B}_{w_i}(\mathbf{O}_{w_i}, r_{w_i})$ and $\mathbb{B}_{w_i}((\alpha_{w_i}, l_{w_i}), r_{w_i})$ are two specific representations of $\mathbb{B}_{w_i}$. Vector concatenation is a reasonable and favorite method to create new vector to carry multiple information (Han et al. 2016; Zeng et al. 2014). The orientation of $\mathbb{B}_{w_i}$'s central point vector, α_{w_i}, is a concatenation of three vector segments: (1) the pre-trained word embedding $\mathbf{w}$, (2) the *PLV* of w_i, (3) a *Spatial Extension Code (SEC)*, as illustrated in Fig. 6.3. The function of *PLV* and *SEC* is to guarantee that all created $\mathscr{N}$-Balls will not contain the origin point of the space (the Principle of Increasing Dimensions). In our case, the concatenated vector contains the information of co-occurrence relations among words (captured by word-embeddings), and the information of the structural information of a word-sense (captured by *PLV*). This structure leads to a novel and powerful method for membership validation (see Sect. 6.5.4 and Chap. 7).

Using Algorithm 1 in Sect. 5.7, we can create a Geometric Connectionist Machine $\mathbf{GCM}_0$ that is able to spatialize unlabeled tree structures. Following (Blackburn 2000), we describe a tree as follows.

Definition A **tree** $\mathfrak{T}$ is a relational structure (T, S) where:

1. T is the set of nodes $\{t_1, t_2, \ldots, t_n\}$.
2. S is the set of node pairs $\{(t_i, t_j)|t_i, t_j \in T\}$.
3. **tree** $\mathfrak{T}$ has a unique root node r, which means that for any other node t there is a unique chain $[r = t'_1, \ldots, t'_i, \ldots, t = t'_w]$ under the condition that neighborhood nodes, t'_j and t'_{j+1}, form a node pair in S, that is, $(t'_j, t'_{j+1}) \in S$.
4. For every non-root node u, there is a unique node v, such that (v, u) in S.
5. For any node x, (x, x) does not exist in S.
6. No other pairs are in S. □

The nodes in T can appear in a dataset. For example, people's names in a family tree can appear in everyday communication, animal names in the taxonomy of zoology can appear in newspaper reports. Associative or co-occurrence relations in corpus can be approximated by connectionist networks in terms of vector embeddings. The cosine value of two vectors discloses the degree of co-occurrence of the corresponding nodes in the corpus. We use $\mathfrak{V}$ as the one-to-one mapping from T to the set of vector embeddings of nodes in T. We write the initialization process as follows.

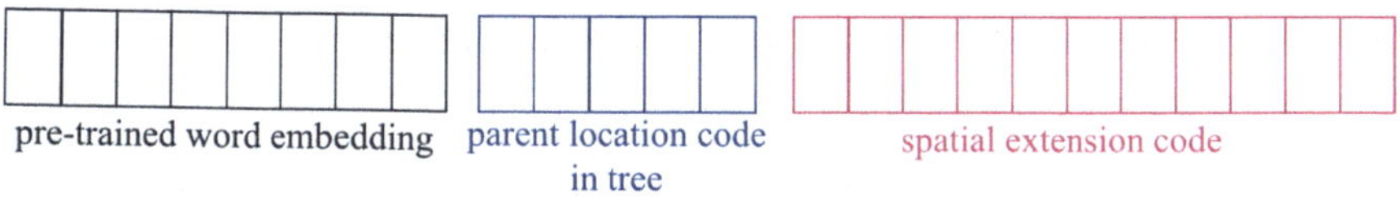

Fig. 6.3 Three components of the center point of an $\mathscr{N}$-Ball embedding

Definition The Θ_0 operator initializes an $\mathscr{N}$-Ball of node $t \in T$ using (1) $\mathfrak{V}$, (2) t's parent location code $\mathbf{plc}(t)$, (3) spatial extension code $\mathbf{se}$, (4) the length of the central point vector d_0, and (5) the radius $r_0 = 0^+$. The vector extension function $\mathfrak{W}$ is structured by $\mathbf{plc}(t)$ and $\mathbf{se}$, $\mathfrak{W}(t) = \mathbf{plc}(t)^\frown\mathbf{se}$.

$$c_t = d_0 \cdot \frac{\mathfrak{V}(t)^\frown\mathfrak{W}(t)}{\|\mathfrak{V}(t)^\frown\mathfrak{W}(t)\|} = d_0 \cdot \frac{\mathfrak{V}(t)^\frown\mathbf{plc}(t)^\frown\mathbf{se}}{\|\mathfrak{V}(t)^\frown\mathbf{plc}(t)^\frown\mathbf{se}\|} \tag{6.1}$$

$$\Theta_0(t, \mathfrak{V}, \mathfrak{W}) = \{p \mid \|p - c_t\| < r_0\} \tag{6.2}$$

□

The set of spatial relations Ξ consists of one relation: *properly contain* (**pContain**). For any $\mathscr{N}$-Ball $\odot_x$ and $\odot_y$, $\mathbf{pContain}(\odot_x, \odot_y)$ means that $\mathscr{N}$-Ball $\odot_x$ *properly contains* $\mathscr{N}$ *-Ball* $\odot_y$. That is, $\Psi : S \to \{\mathbf{pContain}\}$. For any $\mathscr{N}$-Ball $\odot_x$ and $\odot_y$, $\mathbf{DC}(\odot_x, \odot_y)$ means that *$\mathscr{N}$-Ball $\odot_x$ disconnects from $\mathscr{N}$-Ball $\odot_y$*. A grounding (or diagramming) operator $\Theta_1 : \bigodot \times S \times \psi \to \bigodot$ is defined as follows.

1. relations in S are spatially interpreted as follows: *for each $(t_i, t_j) \in S$, it holds that the $\mathscr{N}$-Ball of t_i properly contains the $\mathscr{N}$-Ball of t_j, that is,* $\mathbf{pContain}(\odot_{t_i}, \odot_{t_j})$;
2. for any $(x, y) \in S$, $(x, z) \in S$, it holds that the $\mathscr{N}$-Ball of y disconnects from the $\mathscr{N}$-Ball of z, that is, $\mathbf{DC}(\odot_y, \odot_z)$.

Formally, $\mathbf{pContain}(\odot_x, \odot_y)$ and $\mathbf{DC}(\odot_x, \odot_y)$ can be defined by the connection relation $\mathbf{C}(\odot_x, \odot_y)$ as follows.

Definition $\mathscr{N}$-Ball $\odot_x$ containing $\mathscr{N}$-Ball $\odot_y$ is defined as that for any $\mathscr{N}$-Ball $\odot_z$, if $\odot_z$ connects with $\odot_y$, $\odot_z$ will connect with $\odot_x$.

$$\mathbf{Contain}(\odot_x, \odot_y) \triangleq \forall \odot_z [\mathbf{C}(\odot_z, \odot_y) \to \mathbf{C}(\odot_z, \odot_x)] \tag{6.3}$$

Definition $\mathscr{N}$-Ball $\odot_x$ properly containing $\mathscr{N}$-Ball $\odot_y$ is defined as $\odot_x$ contains $\odot_y$ and $\odot_y$ does not contain $\odot_x$.

$$\mathbf{pContain}(\odot_x, \odot_y) \triangleq \mathbf{Contain}(\odot_x, \odot_y) \wedge \neg\mathbf{Contain}(\odot_y, \odot_x) \tag{6.4}$$

Definition $\mathscr{N}$-Ball $\odot_x$ disconnecting from $\mathscr{N}$-Ball $\odot_y$ is defined as $\odot_x$ does not connect with $\odot_y$.

$$\mathbf{DC}(\odot_x, \odot_y) \triangleq \neg\mathbf{C}(\odot_x, \odot_y) \tag{6.5}$$

Theorem **pContain** *is irreflexive, that is,* $\forall \odot_x [\neg\mathbf{pContain}(\odot_x, \odot_x)]$ □

Theorem **pContain** *is transitive,* $\forall \odot_x, \odot_y, \odot_z [\mathbf{pContain}(\odot_x, \odot_y) \wedge \mathbf{pContain}(\odot_y, \odot_z) \to \mathbf{pContain}(\odot_x, \odot_z)]$ □

Theorem *There is a unique $\mathscr{N}$-Ball $\odot_r$ that properly contains any other $\mathscr{N}$-Ball $\odot_x$.* □

Theorem *For every $\mathscr{N}$-Ball $\odot_x$, if there are two different $\mathscr{N}$-Balls $\odot_p$ and $\odot_q$ such that* **pContain**$(\odot_p, \odot_x)$ *and* **pContain**$(\odot_q, \odot_x)$*, then either $\odot_p$ properly contains $\odot_q$, or $\odot_q$ properly contains $\odot_p$, that is,* **pContain**$(\odot_p, \odot_q) \vee$ **pContain**$(\odot_q, \odot_p)$. □

A spatializing operator Θ spatializes $\mathfrak{S}$ onto $\mathscr{N}$-Balls in $(m+n)$-dimensional space, while preserving $\mathfrak{V}$ and $\mathfrak{W}$.

$$\Theta(\mathfrak{S}, \mathfrak{V}, \mathfrak{W}, \psi) = \Theta_1(\Theta_0(T, \mathfrak{V}, \mathfrak{W}), S, \psi)$$

This Geometric Connectionist Machine ($\mathbf{GCM}_0$) takes $\mathfrak{T} = (T, S)$ and $\mathfrak{V}$ as input, initializes nodes in T by Θ_0, and uses the set of geometric transformations $\mathfrak{G} = \{\mathsf{H}, \mathsf{S}, \mathsf{R}\}$ to implement Θ_1. The outputs of $\mathbf{GCM}_0(\mathfrak{T}, \mathfrak{V})$ are two folded: (1) all constructed $\mathscr{N}$-Balls of nodes in T, and (2) for each constructed $\mathscr{N}$-Ball, the sequence of geometric operations which describes the transformations from its initial state to its final state, $\underline{\Theta_1 \triangleright \mathfrak{G}}(\Theta_0(T, \mathfrak{V}, \mathfrak{W}), S, \psi)$. We use $\underline{\Theta_1 \triangleright \mathfrak{G}}$ to represent the implementation of the abstract function Θ_1 through functions in $\mathfrak{G}$. That is,

$$\mathbf{GCM}_0(\mathfrak{T}, \mathfrak{V}) = (\bigodot, \Theta(\mathfrak{S}, \mathfrak{V}, \mathfrak{W}, \psi)) \tag{6.6}$$

$$= (\bigodot, \underline{\Theta_1 \triangleright \mathfrak{G}}(\Theta_0(T, \mathfrak{V}, \mathfrak{W}), S, \psi)) \tag{6.7}$$

In von Neumann machines, variables can store either data, or address of be another variable. Geometric Connectionist Machines also provide two different ways for addressing $\mathscr{N}$-Balls: The static address of an $\mathscr{N}$-Ball is its center point and the radius, and the dynamic address of an $\mathscr{N}$-Ball is a route description in terms of a sequence of geometric transformations. A route description starts from its initial address of an $\mathscr{N}$-Ball, followed by a sequence of geometric transformations.

6.3 The Experiment with Word-Senses

6.3.1 Input Datasets

We use GloVe 50-dimension word embedding[1] as the pre-trained word embeddings (Pennington et al. 2014), and extract hypernym tree structures of word-senses from WordNet 3.0 (Miller 1995). We focus on nouns and verbs, totaling 291 trees, 54, 310 nodes. The tree with the root entity.n.01 is the largest, totaling 43, 669 nodes. All tree structures can be downloaded at https://figshare.com/articles/glove_6B_tree_input_zip/7607297.

[1] GloVe embeddings can be downloaded at https://nlp.stanford.edu/projects/glove/.

6.3.2 *Output Datasets*

The output is 54, 310 $\mathcal{N}$-Ball embeddings, each embedding consisting of three parts: (1) the normalized center point vector; (2) the length of the center point vector; (3) the radius. All $\mathcal{N}$-Balls are free for the public access at https://figshare.com/articles/glove_6B_50Xball_txt_zip/7607321. The construction process took around 3 hours at Mac Pro platform with Intel Core i7 processor (2,5 GHz).

6.4 Evaluation

6.4.1 *Tree Structure Is Precisely Spatialized*

We use python scripts to check whether each child-parent relation is correctly encoded by the inclusion relation and whether each sibling relation is correctly encoded by the disconnectedness relation in the corresponding $\mathcal{N}$-Balls. Python scripts returns no error message. That is, the geometric approach has successfully spatialized a large tree structure onto word-embedding at the zero energy cost, which satisfies our strict criteria set in Chap. 4.

6.4.2 *Pre-trained Word-Embeddings Are Well Preserved*

The aim of this experiment is to examine the effect of the geometric transformation process to the pre-trained word-embedding. We examine the standard deviation (std) of pre-trained word-embeddings in $\mathcal{N}$-Ball representations. The less the std is, the better it is preserved. The trained $\mathcal{N}$-Ball representations have 32,503 word-stems. For each word-stem, we extract word-embedding parts from the central point vector of $\mathcal{N}$-Ball representations, normalize them, and compute standard deviation. The maximum std is 0.2838. There are 94 stds greater than 0.2, 198 stds in the range of $(0.1, 0.2]$, 25 stds in the range of $(10^{-12}, 0.1]$, 4124 stds in the range of $(0, 10^{-12}]$, 28,062 stds equals 0. With this statistics we conclude that only a very tiny portion (0.898%) of pre-trained word-embeddings have a small change (std $\in (0.1, 0.2838]$).

6.4.3 *Consistency to Benchmark Tests*

The quality of word-embedding is evaluated by computing the consistency (Spearman's correlation) between human-judged word relations and vector-based similarity relations. The standard datasets in the literature are WordSim353 dataset, and Stanford's Contextual Word Similarities (SCWS) dataset. WordSim353 consists of 353 pairs of words, each pair is associated with a human-judged value about the co-relation between the two words (Finkelstein et al. 2001); SCWS contains 2003 word pairs each with 10 human judgments on similarity (Huang et al. 2012).

6.4.3.1 Dataset

Unfortunately, both datasets cannot be directly used within our experiment setting due to several reasons as follows.

- Words whose word senses having neither hypernym, nor hyponym in WordNet 3.0, e.g. holy
- Words whose word senses having different word stems in WordNet 3.0, e.g. laboratory, midday, graveyard, percent, zoo, woodland, FBI, personnel, media, CD, wizard, rooster, medal, hypertension, valor, lad, OPEC
- Words having no word senses in WordNet 3.0, e.g. Maradona
- Words whose word senses using its basic form as word stems in WordNet 3.0, e.g. clothes, troops, earning, fighting, children

After removing all the missing words, we have 318 paired words from WordSim353 and 1719 pairs from SCWS dataset for the evaluation.

6.4.3.2 Result

Using the pre-trained word-embedding and the word-embedding segments in $\mathcal{N}$-Ball embeddings, we get the same Spearman's correlation value in all testing cases,

Table 6.1 Word embedding part $WE_{\mathcal{N}-Ball}$ in $\mathcal{N}$-Balls produce the same Spearman's correlation as the pertained word-embedding WE_{GloVe}

	$\rho : H \sim WE_{GloVe}$	$\rho : H \sim WE_{\mathcal{N}-Ball}$
WordSim318	0.7608233126890	0.7608233126890
SCWS1719	−0.9417662761569	−0.9417662761569
	−0.9470593058763	−0.9470593058763
	−0.9331354644654	−0.9331354644654
	−0.9253900797835	−0.9253900797835
	−0.9502155219153	−0.9502155219153
	−0.9366014705195	−0.9366014705195
	−0.9365441574709	−0.9365441574709
	−0.9450659844806	−0.9450659844806
	−0.9344177076472	−0.9344177076472
	−0.9455129213692	−0.9455129213692

as listed in Table 6.1. We conclude that despite of shift and rotation transformations, the semantic value of the word-embedding part is well preserved in geometric transformation embeddings.[2]

6.5 Experiments with Geometric Connectionist Machines

To better understand geometric transformation embeddings, we perform a number of experiments and compare results with pre-trained GloVe and ConceptNet word-embeddings.

6.5.1 *Similarity Measurement and Comparison*

In the literature of representation learning, similarity is normally measured by the cosine value of two vectors (Mikolov et al. 2013a). In $\mathcal{N}$-Ball representation, the similarity between two balls can be approximated by the cosine value of their center points. Formally, given two $\mathcal{N}$-Balls $\mathbb{B}_{w_i}((\alpha_{w_i}, l_{w_i}), r_{w_i})$ and $\mathbb{B}_{v_j}((\alpha_{v_j}, l_{v_j}), r_{v_j})$, their cosine similarity can be defined as $\cos(\alpha_{w_i}, \alpha_{v_j})$. Intuitively, two word-senses are similar, if they are siblings, $\textbf{SIBLING}(\mathbb{B}_{w_i}, \mathbb{B}_{w_j})$ is true. We introduce this typed cosine similarity as follow.

$$Sim_\tau(w_i, v_j) \triangleq \begin{cases} \cos(\alpha_{w_i}, \alpha_{v_j}) & \textbf{SIBLING}(\mathbb{B}_{w_i}, \mathbb{B}_{v_j}) \\ -1 & \text{otherwise} \end{cases}$$

6.5.2 *Nearest Neighbors*

The first evaluation is to inspect nearest neighbors of some polysemous words, to see whether their nearest neighbors are meaningful. For example, Berlin can denote a city (berlin.n.01), or a family name (berlin.n.02). We compare results with GloVe embeddings and ConceptNet embeddings (Speer et al. 2017). ConceptNet embeddings use *retrofitting* approach (Faruqui et al. 2015) to integrate word-embedding information with knowledge-graph embeddings. That is, a ConceptNet embedding $\mathbf{p}$ of the concept c shall be close to its word-embedding of c and its neighbor embedding $\mathbf{q_i}$ of the concept c_i, in which c directly relates with c_i in the knowledge graph.

The top-5 nearest neighbors (based on Sim_τ) of beijing.n.01, berlin.n.01, and berlin.n.02 are listed in Table 6.2. We see similarities measured by $\mathcal{N}$-Ball embeddings are almost 1, much higher than that measured by GloVe or ConceptNet embed-

[2]We need to further investigate why Spearman's correlations are less than 0 for SCWS1719, It is beyond the topic of this experiment.

Table 6.2 Top-5 nearest neighbors measured by Sim_τ and cos, respectively

Word-sense 1	Word-sense 2	Sim_τ	cos GloVe stem	cos ConceptNet stem
Beijing.n.01	London.n.01	0.99999997	0.47	0.31
	Paris.n.01	0.99999997	0.46	0.33
	Atlanta.n.01	0.99999992	0.27	0.27
	Boston.n.01	0.99999985	0.29	0.26
	Baghdad.n.01	0.99999975	0.40	0.25
Berlin.n.01	Rome.n.01	0.99999993	0.68	0.25
	Madrid.n.01	0.99999997	0.47	0.23
	Toronto.n.01	0.99999997	0.46	0.21
	Columbia.n.03	0.99999992	0.40	0.16
	Sydney.n.01	0.99999987	0.52	0.27
Berlin.n.02	Simon.n.02	0.99999996	0.34	0.04
	Williams.n.01	0.99999996	0.24	0.02
	Foster.n.01	0.99999992	0.13	0.02
	Dylan.n.01	0.99999984	0.10	0.09
	Mccartney.n.01	0.99999974	0.23	0.04

dings. Using GloVe embeddings, the top nearest neighbors of beijing are: china, taiwan, seoul, taipei, chinese, among which only seoul and taipei are cities; the top nearest neighbors of berlin look much better: vienna, warsaw, munich, prague, germany, moscow, hamburg, bonn, copenhagen, cologne, among which germany is a country. However, this list is highly biased: the word-sense of family name does not appear. Using pre-trained ConceptNet word-embedding vectors,[3] we identified the top-5 nearest neighbors of beijing as follow: peiping, chung_tu, zhongdu, in_china, peking, among which peiping and peking are synonyms of beijing, chung_tu and zhongdu are synonyms denoting a district of beijing in the history, while in_china is not city name; the top-10 nearest neighbors of berlin are: berlinese, irving_berlin, east_berlin, ####_ndash_####, frankfurt_der_oder, german_empire, west_berlin, impero_tedesco, frankfurt_on_main, germany, among which berlinese refers to residents of the city Berlin, irving_berlin is a person's name, impero_tedesco refers to German Empire (1871–1918) (Table 6.3).

We conclude that the cosine similarity using $\mathscr{N}$-Ball embeddings is more meaningful than those measured by GloVe or ConceptNet word-embeddings.

[3] https://conceptnet.s3.amazonaws.com/downloads/2017/numberbatch/numberbatch-en-17.06.txt.gz.

Table 6.3 Nearest neighbors using GloVe embeddings, and ConceptNet embeddings

Word	Top neighbor using GloVe embeddings	Top neighbor using ConceptNet embeddings
Beijing	*China*, *Taiwan*, Seoul, Taipei	Peiping, Chung_tu, Zhongdu, in_China, Peking
	Chinese, Shanghai, *Korea*	
	Mainland, *Hong*, *Wen*, Kong	
Berlin	Vienna, Warsaw, Munich	Berlinese, *Irving_Berlin*, *East_Berlin*
	Prague, *Germany*, Moscow	*####_ndash_####*, Frankfurt_Der_Oder
	Hamburg, Bonn, Copenhagen	German_Empire, West_Berlin, Impero_Tedesco
		Frankfurt_am_main, Germany

6.5.3 *Representing Structures Beyond Word-Embeddings*

We inspected how category structure is encoded in $\mathcal{N}$-Ball embeddings, and found that any word-sense w that produces a negative value of $f_{\mathbf{P}}(\mathbb{B}_u, \mathbb{B}_w)$ is a hypernym (upper category) of u (listed in Table 6.4). The $\mathbb{B}_w$ producing the maximum

Table 6.4 Upper-level categories based on $f_{\mathbf{P}}$. $-_k$ represents the kth largest negative value of $f_{\mathbf{P}}$

Word-sense 1	Word-sense 2	$f_{\mathbf{P}}$
Beijing.n.01	City.n.01	$-_1$
Berlin.n.01	Municipality.n.01	$-_2$
Prague.n.01	Region.n.03	$-_3$
Paris.n.01	Location.n.01	$-_4$
Madrid.n.01	Object.n.01	$-_5$
Berlin.n.02	Songwriter.n.01	$-_1$
Simon.n.02	Composer.n.01	$-_2$
Williams.n.01	Musician.n.02	$-_3$
Foster.n.01	Artist.n.01	$-_4$
Dylan.n.01	Creator.n.02	$-_5$
Mccartney.n.01	Person.n.01	$-_6$
Peirce.n.02	Philosopher.n.01	$-_1$
Nietzsche.n.01	Scholar.n.01	$-_2$
Hegel.n.01	Intellectual.n.01	$-_3$
Descartes.n.01	Person.n.01	$-_4$

negative value is the Direct Upper Category of u, written as $\mathbf{DUC}(\mathbb{B}_u)$. For example, $f_{\mathbf{P}}(\mathbb{B}_{beijing.n.01}, \mathbb{B}_{city.n.01})$ is the maximum value among the set $\{f_{\mathbf{P}}(\mathbb{B}_{beijing.n.01}, \mathbb{B}_w) | f_{\mathbf{P}}(\mathbb{B}_{beijing.n.01}, \mathbb{B}_w) < 0\}$.

$$\mathbf{DUC}(\mathbb{B}_u) = \arg\max_{\mathbb{B}_w}\{f_{\mathbf{P}}(\mathbb{B}_{beijing.n.01}, \mathbb{B}_w) | f_{\mathbf{P}}(\mathbb{B}_{beijing.n.01}, \mathbb{B}_w) \leq 0\}$$

The symbolic relation of Direct Upper Category is precisely computed as inclusion relations among balls in a continuous space. Neither GloVe embeddings, nor ConceptNet embeddings can do this.

6.5.4 *Word-Sense Validation as Way-Finding*

Our feet carry us from place to place along paths, just as our minds carry us from idea to idea along paths.

— Barbara Tversky

The $\mathbf{GCM}_0$ precisely creates $\mathcal{N}$-Ball configurations of word-senses by synergistically integrating their co-occurrence information, and symbolic tree structured category information. For each created $\mathcal{N}$-Ball, $\mathbf{GCM}_0$ provides two pieces of information as follows.

- static information: its finial location and the radius
- dynamic information: the sequence of geometric transformations that transform an $\mathcal{N}$-Ball from its initial status to the final status.

The dynamic information can be likened as a route instruction that tells a baby's home address starting from the hospital address where it is born. If we already have the route instruction of its siblings, we can use it to send the new born baby home. In this experiment, we will use this idea to validate the category of an unknown word that appears in corpus. For example, when we read a text *Solingen has long been renowned for the manufacturing of fine swords, knives, scissors and razors ...*, we wonder whether *Solingen* is a *city* or a *person*? Let us suppose it is a city, and initialize its $\mathcal{N}$-Ball by using Parent Location Code of a known member of city. e.g. *Berlin*. We have the route instruction to send *Berlin* from its birth hospital into the $\mathcal{N}$-Ball of *city*. So, we use this route to send newly born $\mathcal{N}$-Ball of *Solingen*. If it is finally located inside $\mathcal{N}$-Ball of *city*, we predict that *Solingen* is a *city*, otherwise not.

6.5.4.1 New Datasets

We create a series of new datasets for membership validation as follows: From 54,310 word-senses (entities), we randomly selected 100 word-senses, with the condition that each of them has at least 10 child nodes. We randomly select $p\%$, $p \in [1, 5, 10, 20, \ldots, 80, 90]$, from child nodes as training set, the rest $1 - p\%$ as testing set. Meanwhile, we randomly select 100 word-senses from the knowledge graph, which are neither hypernyms nor hyponyms of the tested entity, and another 100 words from corpus, whose word-senses are not in the knowledge base. For example, if we choose 1% of child nodes of city.n.01 and the rest 99% as testing. We create a file with the name membershipPredictionTask.txt10_1 (10_1 means the total number of child nodes is at least 10, 1% is used as training), with a line as follows: city.n.01#linz.n.01 cremona.n.01 winchester.n.01 fargo.n.01 philippi.n.01 atlanta.n.01 medan.n.01#bologna.n.01 medina.n.0...#stick.v.12 cartagena.n.01 ...#boozed optima tirpitz per ... in the format of 'superordinate#known subordinates#unknown subordinates#100 incorrect known word-senses#100 words not in knowledge graph', see Appendix A for detail. All membership validation tasks are available for public access https://figshare.com/articles/membership_prediction_tasks_zip/7645085.

6.5.4.2 The Geometric Transformation Approach

We describe the task formally as follows: Given a pre-trained d-dimension word-embedding $\mathscr{E}_N : W_N \rightarrow R^d$, W_N is the set of word, and the hypernym structure $\mathscr{T}_K$ of word-senses. Let $x.p.n$ be an unknown word-sense, that is, $x \in W_N$ and $x.p.n \notin \mathscr{T}_K$. Let $h \in W_K$, can we validate that $x.p.n$ is a member of h?

The geometric procedure is described as follows: to validate whether $x.p.n$ is a member of $h \in W_K$, we get the full hypernym path of h: $[h_0, h_1, \ldots, h_n, h]$, as well as the parent location vector PLV_h of h's known member in $\mathscr{T}_K$; then, we construct an geometric transformation representation for the tree T_h only with nodes $h_0, h_1, \ldots, h_n, h$ and all the known children of h. The construction process produces a transformation log file, $trans_{his}$, which records all the transformations in sequence. Thirdly, we initialize the ball of x with Parent Location Vector PLV_h , and apply the same transformations in $trans_{his}$ in sequence. If we find the x ball is located inside of the ball h, we can say that h is the hyponym of x.

6.5.4.3 Evaluation and Analysis

The experiment result shows that the geometric transformation method is very precise and robust:

- The precision is always 100%, even we only select 1% of the children as training set.
- The recall increases, with small fluctuations, if we select more percentage of the children from the training set.
- The method is very robust. If we only select 1% of the children as training set, the recall of prediction reaches 68.6% with the F1 score of 81.4%; if we select 5% as

Table 6.5 The table of recall with different parameters. "% in Training" represents how much percentage is used as training data; "Ratio × Radius" represents the radius r being enlarged by multiplying a ratio

% in Training	Ratio × Radius										
	Ratio = 1.0	Ratio = 1.1	Ratio = 1.2	Ratio = 1.3	Ratio = 1.4	Ratio = 1.5	Ratio = 1.6	Ratio = 1.7	Ratio = 1.8	Ratio = 1.9	Ratio = 2.0
1	0.68	0.82	0.90	0.95	0.98	0.99	0.99	0.99	0.99	0.99	0.99
5	0.79	0.89	0.94	0.97	0.98	0.99	0.99	0.99	1.0	1.0	1.0
10	0.84	0.92	0.97	0.99	0.99	0.99	0.99	1.0	1.0	1.0	1.0
20	0.92	0.97	0.99	0.99	0.99	1.0	1.0	1.0	1.0	1.0	1.0
30	0.94	0.98	0.99	0.99	1.0	1.0	1.0	1.0	1.0	1.0	1.0
40	0.96	0.99	0.99	0.99	1.0	1.0	1.0	1.0	1.0	1.0	1.0
50	0.96	0.98	0.99	0.99	1.0	1.0	1.0	1.0	1.0	1.0	1.0
60	0.96	0.99	0.99	1.0	1.0	1.0	1.0	1.0	1.0	1.0	1.0
70	0.98	0.99	0.99	1.0	1.0	1.0	1.0	1.0	1.0	1.0	1.0
80	0.97	0.99	1.0	1.0	1.0	1.0	1.0	1.0	1.0	1.0	1.0
90	0.96	0.99	1.0	1.0	1.0	1.0	1.0	1.0	1.0	1.0	1.0

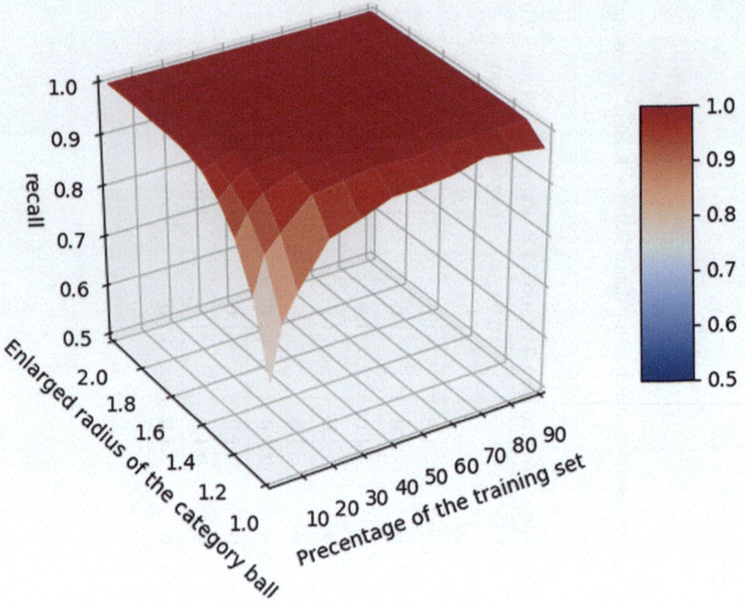

Fig. 6.4 Recall increases rapidly with the increasing of the radius

training set, the recall of prediction reaches 79.1% withe the F1 score of 88.3%; if we select 50% as training set, the recall of prediction reaches 96.4% withe the F1 score of 98.2%, as listed in Table 6.5 and illustrated in Figs. 6.4 and 6.5a.

To increase values of the recall, we enlarge r_h. Using the same datasets, we repeat the above experiment by multiplying r_h with ratio $\beta = [1.1, 1.2, 1.3, 1.4, 1.5, 1.6, 1.7, 1.8, 1.9, 2.0]$. Recall values increase rapidly, without sacrificing precision values, as listed in Table 6.5, and illustrated in Fig. 6.4. For example, if we set $\beta = 1.1$, the overall recall increases to 82.15% from 68.57% for the dataset with 1% of the known members. Theoretically, this method shall not guarantee 100% recall. Two features are observed as follows.

- the mode of recall is always 1.0, even only 1% of the children are selected as training set. That is, in most case, the boundary we draw does not exclude true members, as shown in Fig. 6.5b.
- the mean of recall will increase, with small fluctuations, if more percentage of the children are selected as training data.

The population standard deviation (pstdev) will decrease, with small fluctuation, if we select more percentage of child nodes as training data. However, there is a relatively big jump, when 90% of the child nodes are selected as training data. The reason lies in the fact that if only a small number (e.g. one) of testing members and these members are excluded outside the ball of their direct upper-category, the recall

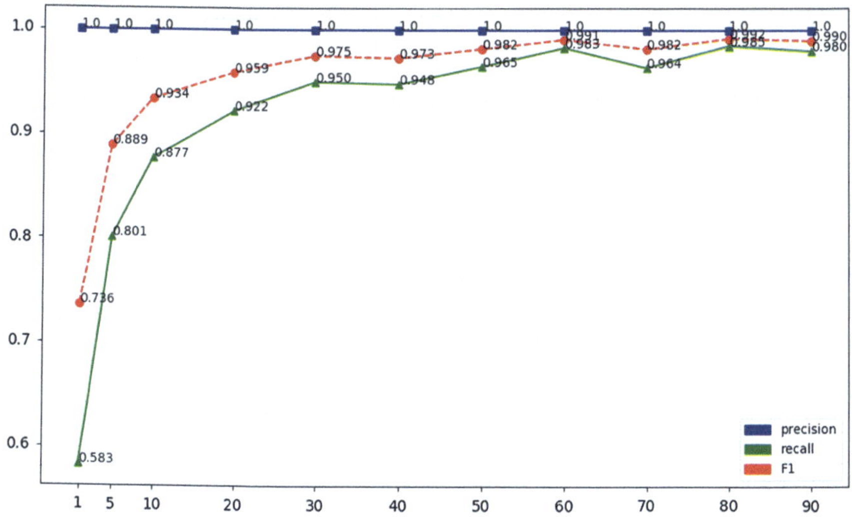

(a) Precision, recall, F1 score of hypernym prediction with 1%, 5%, ... of known members

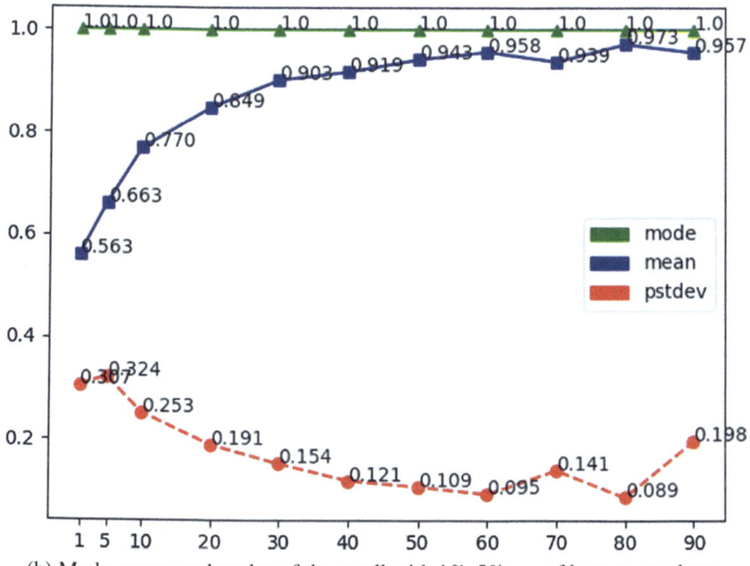

(b) Mode, mean, and pstdev of the recall with 1%, 5%, ... of known members

Fig. 6.5 Precision and recall without margin extension

will be dropped to 0%, which increases the value of pstdev. All experiment results can be downloaded at https://figshare.com/articles/membership_prediction_results_zip/7607351.

6.6 Summary

We successfully spatialized a large tree structure onto word-embeddings by utilizing Geometric Connectionist Machine. Experiment results are surprisingly great, especially, the precision of validating unknown word-senses, which needs further investigation.

References

Blackburn, P. (2000). Representation, reasoning, and relational structures: A hybrid logic manifesto. *Logic Journal of the IGPL*, *8*(3), 339–625.

Faruqui, M., Dodge, J., Jauhar, S. K., Dyer, C., Hovy, E., & Smith, N. A. (2015). Retrofitting word vectors to semantic lexicons. In *Proceedings of the 2015 Conference of the North American Chapter of the Association for Computational Linguistics: Human Language Technologies*, pp. 1606–1615. ACL.

Finkelstein, L., Gabrilovich, E., Matias, Y., Rivlin, E., Solan, Z., Wolfman, G., et al. (2001). Placing search in context: The concept revisited. WWW, 406–414.

Han, X., Liu, Z., & Sun, M. (2016). Joint representation learning of text and knowledge for knowledge graph completion. CoRR, arXiv:abs/1611.04125.

Huang, E. H., Socher, R., Manning, C. D., & Ng, A. Y. (2012). Improving word representations via global context and multiple word prototypes. In *ACL'12: Long Papers - Volume 1*, pp. 873–882, Stroudsburg, PA, USA. Association for Computational Linguistics.

Mikolov, T., Chen, K., Corrado, G., & Dean, J. (2013). Efficient estimation of word representations in vector space. CoRR, arXiv:abs/1301.3781.

Miller, G. A. (1995). Wordnet: A lexical database for english. *Communication ACM*, *38*(11), 39–41.

Pennington, J., Socher, R., & Manning, C. D. (2014). GloVe: Global vectors for word representation. In *EMNLP'14*, pp. 1532–1543.

Speer, R., Chin, J., & Havasi, C. (2017). Conceptnet 5.5: An open multilingual graph of general knowledge. In *Proceedings of the Thirty-First AAAI Conference on Artificial Intelligence*, pp. 4444–4451, February 4–9, 2017, San Francisco, California, USA.

Zeng, D., Liu, K., Lai, S., Zhou, G., & Zhao, J. (2014). Relation classification via convolutional deep neural network. *COLING*, pp. 2335–2344.

Chapter 7
Geometric Connectionist Machines for Triple Classification

A creature didn't think in order to move; it just moved, and by moving it discovered the world that then formed the content of its thoughts.

— Larissa MacFarquhar "The mind-expanding ideas of Andy Clark" The New Yorker

This chapter is the continued discussion on the last experiment in Chap. 6—Under what condition, can the precision for the Task of Membership-Validation reach 100%? We will create a new type of Geometric Connectionist Machines for Triple Classification task in Knowledge Graph reasoning. Our key question is: How shall we spatialize labeled tree structures onto vector embeddings?

7.1 Introduction

Knowledge Graphs represent truth knowledge in the Triple form of *(head, relation, tail)*, shortened for (h, r, t). For example, *(berlin.n.01, is_a, city.n.01)* is a Triple. Knowledge Graphs (Miller 1995; Suchanek et al. 2007; Auer et al. 2007; Bollacker et al. 2008) are very useful for AI applications (Manning et al. 2008b; Socher et al. 2013; Bordes et al. 2013; Wang et al. 2014; Wang and Li 2016). However, Knowledge Graphs normally suffer from incompleteness. One research topic in AI is to predict the missing part of a Knowledge Graph. *Triple Classification* is to determine the truth value, or the degree of truth, of an unknown Triple.

T. Dong, *A Geometric Approach to the Unification of Symbolic Structures and Neural Networks*, Studies in Computational Intelligence 910,
https://doi.org/10.1007/978-3-030-56275-5_7

Inspired by the training method used in word-embedding, Bordes et al. (2013) used margin-based score function to train vector representations of Triples (h, r, t), such that the vector triangle equation $\mathbf{t} \approx \mathbf{h} + \mathbf{r}$ should hold for positive samples and that $\mathbf{t}$ should differ from $\mathbf{h} + \mathbf{r}$ with a margin for negative samples. An appealing feature of this method is that discrete entities can be approximated by their corresponding vectors in the continuous space. However, it is not difficult to understand that precise vector representation does not exist for 1-to-N relation: Let $\mathbf{p}$ be the precise vector representation for the *part-of* relation, *(head, part-of, body)* and *(toe, part-of, body)* be two assertions. Ideally, we shall have $\mathbf{head} + \mathbf{p} = \mathbf{body}$ and $\mathbf{toe} + \mathbf{p} = \mathbf{body}$, which leads to $\mathbf{head} = \mathbf{toe} = \mathbf{body} - \mathbf{p}$. We observe that *part-of* relation can also be precisely and implicitly represented between regions, instead of vectors–regions representing *head* and *toe* are located inside the region representing the *body* (see Sect. 3.4 for more discussion). This suggests a region-based approach for Triple Classification.

We encounter two challenges: (1) Knowledge Graphs normally consist of Triples with multiple relations, e.g., (*beijing, isa, city*) and (*inner_city, part_of, city*). Both *beijing* ball and *inner_city* ball shall be located inside *city* ball. We need to distinguish the *isa* relation from the *part_of* relation. (2) It can happen that entities in the testing Triple do not exist in Knowledge Graph.

7.1.1 Subspaces for Multiple Relations

Our solution is to introduce *isa* subspace and *part_of* subspace into *city* $\mathcal{N}$-Ball. *Beijing* $\mathcal{N}$-Ball is located inside the *isa* subspace, while *inner_city* $\mathcal{N}$-Ball is inside the *part_of* subspace. Subspaces are disconnected from each other. Given Triple (h, rel, t), the rel-subspace is constructed within $\mathbb{B}_h(\mathbf{O}_h, r_h)$, and is also an $\mathcal{N}$-Ball, written as $\mathbb{B}_{h_{rel}}(\mathbf{O}_{h_{rel}}, r_{h_{rel}})$. For each tail x such that (h, rel, x), $\mathbb{B}_x(\mathbf{O}_x, r_x)$ is inside the subspace $\mathbb{B}_{h_{rel}}(\mathbf{O}_{h_{rel}}, r_{h_{rel}})$, as illustrated in Fig. 7.1.

7.1.2 Enriching Knowledge Graph Embedding with Text Information

Representation learning only based on Triples of a Knowledge Graph may have difficulty in validating Triples having unknown entities. This limitation can be approached by introducing text information into entity embedding. For example, entities can be aligned with their Wikipedia anchors Wang et al. (2014); entities can be associated with their descriptions (Zhong et al. 2015; Zhang et al. 2015; Xie et al. 2016). Experiment results show that integrating texts into learning representation improves the quality of graph embedding (Faruqui et al. 2015; Wang and Li 2016; Han et al. 2016; Speer et al. 2017). We use TEKE entity embeddings Wang and Li (2016) as pre-trained entity-embeddings.

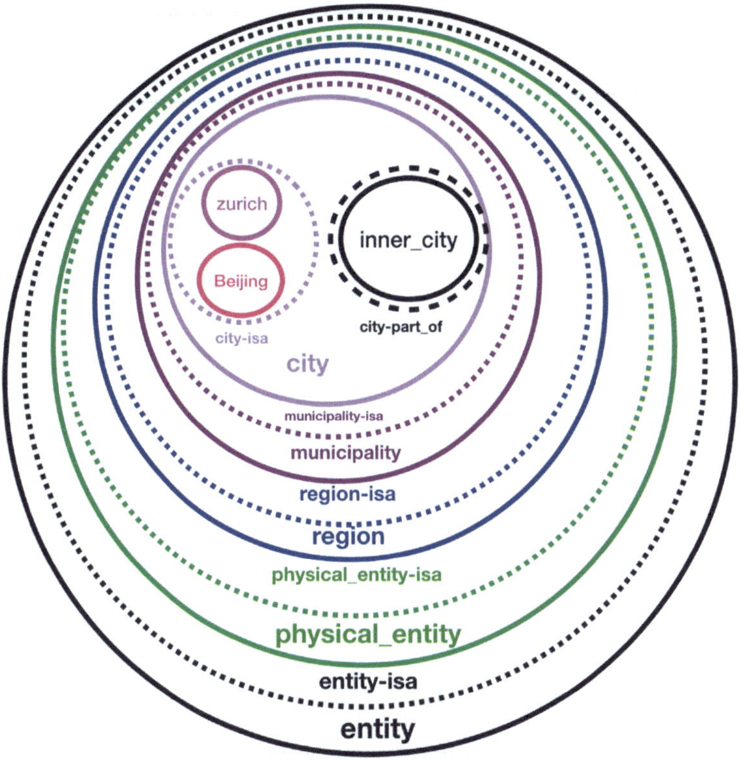

Fig. 7.1 Subspaces inside city ball

7.2 Geometric Connectionist Machines for Triple Classification

Given a new Triple (h, r, t_x), we will construct a geometric model $\mathcal{M}(h, r)$ from the Knowledge Graph that validates whether (h, r, t_x) is true or false. This is realized by inspecting whether the constructed $\mathcal{N}$-Ball of t_x is inside the r-subspace of h as follows: Given tails $t_1, t_2, \ldots, t_n$ of head h with relation r, and a typing chain of h: $h = h_0, h_1, h_2, \ldots, h_k$, we design a geometrical process to construct $\mathcal{N}$-Balls of ts, namely, t_1, ..., t_n, and hs, namely, h_0, ..., h_k, such that any two $\mathcal{N}$-Balls of ts are disconnected, t's $\mathcal{N}$-Ball is inside r-subspace of h's $\mathcal{N}$-Ball, and $\mathcal{N}$-Ball of h_{i-1} is contained by *is-a* subspace of h_i's $\mathcal{N}$-Ball. To determine whether a new entity t_x is a tail of head h with relation r, we initialize the $\mathcal{N}$-Ball of t_x as the $\mathcal{N}$-Ball of its sibling in the training set,[1] and apply the same geometric process to t_x.

[1] That is to say, suppose that (h, r, t_x) holds and t_x has a sibling t_0 in the training dataset such that (h, r, t_0), we initialize the $\mathcal{N}$-Ball of t_x the way we initialize the $\mathcal{N}$-Ball of t_0.

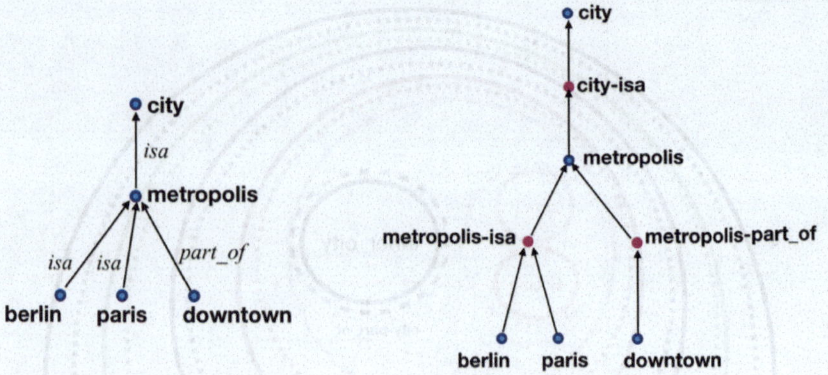

(a) A labeled tree whose edges can either be labeled with *isa* relation or with *part_of* relation

(b) Labels are transformed into nodes (*label nodes*)

Fig. 7.2 Transforming a labeled tree into an unlabeled tree

If the $\mathcal{N}$-Ball of t_x is contained by the r-subspace of $\mathcal{N}$-Ball of h, (h, r, t_x) will be classified as positive.

Triples can be diagrammed as labeled pairs, as illustrated in Fig. 7.2a. We transform a labeled pair into two unlabeled pairs. For example, (city, isa, metropolis) is transformed into (city, city-isa) and (city-isa, metropolis), as illustrated in Fig. 7.2b.

We use depth-first recursive procedure to construct subspaces, as illustrated in Algorithm 1. Sibling subspaces are mutually disconnected. The $\mathcal{N}$-Ball of h is a ball covering of all its r-subspaces. This is the second type of Geometric Connectionist Machine, $\mathbf{GCM}_1$.

7.2.1 *Structure of the Orientation of the Central Point Vector*

The structure of the orientation of the central point consists of four pieces of vector information, as illustrated in Fig. 7.3a.

- The vector embedding of entity e;
- The length p of the type chain of e is encoded by a layer vector (a simple Parent Location Code), whose first p elements are '1', followed by '0'. For example, the layer vector of entity.n.01 is [0,0,0,0,0], the layer vector of location.n.01 is [1,0,0,0,0], the layer vector of xian.n.01 is [1,1,1,1,0], see Fig. 7.3b. Siblings have the same layer vector;
- Different relations to e's parent are encoded by different vectors, which are called *subspace vectors*;
- Spatial extension code.

Algorithm 1: construct r-subspace of $\mathcal{N}$-Ball embeddings

```
input  : known tails: (h, r, t1)...(h, r, tM);
         type chain: tChains =[h, h1, ..., hN];
         pre-trained entity-embeddings: EV
output: r-subspace of 𝒩-Ball of head h: ballHr; geometric transformation history: tranHis
tranHis = [] // initialize 𝒩-Balls of [t1, ...,tM]
foreach ele ∈ [t1, ...,tM] do
 | bTails[ele]=init_nball(ele, EV)
end
// make 𝒩-Balls of [t1, ...,tM] mutually be disconnected using
   three geometric transformations
bTails, tranHis =disconnect(bTails, tranHis) ballHr =init_nball(h, EV) // for each
   member in bTails, create 𝒩-Ball of r-space of h to contain,
   using geometric transformations
foreach ele ∈ bTails do
 | bHr [ele ], tranHis =contain(ballHr, ele, tranHis)
end
while tChains not empty do
 |  ele = pop(tChains, 0) if ele is h then
 |  |  // create the minimal cover of bHr as the r-subspace
 |  |     inside h's 𝒩-Ball
 |  |  ballHr =mini_cover_ball(bHr) ele0 = h
 |  else
 |  |  bTs [ele ]=init_nball(ele, EV) bTs [ele ], tranHis
 |  |   =contain(bTs[ele], ele0, tranHis) ele0 = ele
 |  end
 |  return ballHr, tranHis
end
```

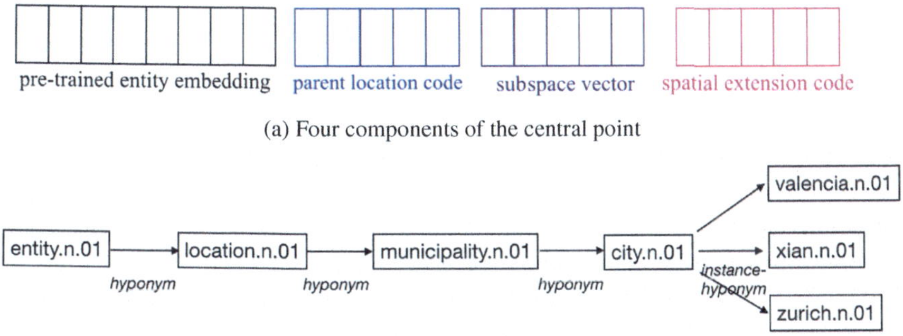

(a) Four components of the central point

(b) A type chain for three concrete cities

Fig. 7.3 The structure of central point vector and a labeled tree

7.2.2 Geometric Connectionist Machine to Spatialize Labeled Tree Structures

The easiest way to create Geometric Connectionist Machine to spatialize labeled trees is to transform a labeled tree structure into a new unlabeled tree structure.

Definition A *labeled tree* $\mathfrak{L}$ is a relational structure $(T \cup L, S_L)$ where:

1. L, the set of labels
2. S_L, the set of triple (x, l, y) such that $x, y \in T$ and $l \in L$
3. T, the set of nodes, contains a unique $r \in T$ (called the $root$) such that r is the predecessor of t, that is, $\forall t \in T$, there is a chain

$$(r, l_0, u_1), (u_1, l_1, u_2), \ldots, (u_k, l_k, t) \tag{7.1}$$

in which $l_i \in L, u_i \in T$.
4. Every element of T distincting from r has a unique predecessor, that is, for every $t \neq r$, there is a unique t' such that $(t', l, t) \in S_L$.
5. For all $t \in T$, there is no chain such that

$$(t, l_0, u_1), (u_1, l_1, u_2), \ldots, (u_k, l_k, t) \tag{7.2}$$

in which $l_i \in L, u_i \in T$
6. No other pairs are in S_L. □

Definition The set of *label nodes*, L_T, is the set of new nodes, whose member is constructed by concatenating the first two elements in S_L.

$$L_T = \{x\text{-}l | (x, l, y) \in S_L\} \tag{7.3}$$

□

We define the *Abbeschriftung* operation abb to transform a labeled tree $\mathfrak{L}$ to an equivalent unlabeled tree $\mathfrak{T}$, as follows.

Definition Given a *labeled tree* $\mathfrak{L} = (T \cup L, S_L)$. The *Abbeschriftung* operation abb transforms $\mathfrak{L}$ to its equivalent unlabeled tree $\mathfrak{T}$.

$$\mathsf{abb}(\mathfrak{L}) = \mathfrak{T} = (\mathscr{T}, \mathscr{S}) \tag{7.4}$$

$$\mathscr{T} = T \cup L_T \tag{7.5}$$

$$\mathscr{S} = \{(x, x\text{-}l), (x\text{-}l, y) | \forall (x, l, y) \in S_L\} \tag{7.6}$$

□

We need to show that $\mathsf{abb}(\mathfrak{L}) = (\mathscr{T}, \mathscr{S})$ is indeed a tree structure defined in Definition 6.2.1 and that labeled tree $\mathfrak{L}$ is equivalent to unlabeled tree $\mathfrak{T}$. Fortunately, both are obvious.

The Geometric Connectionist Machine that spatializes labeled tree $\mathbf{GCM}_1$ takes $\mathfrak{L} = (T, L, S_L)$ and $\mathfrak{V}$ as inputs, in which $\mathfrak{V}$ is the one-to-one mapping from T to the set of vectorial embeddings of T's members. Nodes in L_T are out of the domain of $\mathfrak{V}$. We update the initialization process as follows.

Definition The promotion operator Θ_0 initializes an $\mathscr{N}$-Ball for node $t \in \mathscr{T}$ using $\mathfrak{V}$, t's parent location code $\mathbf{plc}(t)$, subspace vector $\mathbf{ss}(l)$ in which $l \in L$, spatial extension code $\mathbf{se}$. The vector extension function $\mathfrak{W}$ is structured by $\mathbf{plc}(t)$, $\mathbf{ss}(l)$, and $\mathbf{se}$, $\mathfrak{W}(t) = \mathbf{plc}(t)^\frown\mathbf{ss}(l)^\frown\mathbf{se}$, if $l \in L$. d_0 and $r_0 = 0^+$ are initial values of the length of the central point and the radius, respectively.

$$c_t = d_0 \cdot \frac{\mathfrak{V}(t)^\frown\mathfrak{W}(t)}{\|\mathfrak{V}(t)^\frown\mathfrak{W}(t)\|} = \begin{cases} d_0 \cdot \frac{\mathfrak{V}(t)^\frown\mathbf{plc}(t)^\frown\mathbf{ss}(l)^\frown\mathbf{se}}{\|\mathfrak{V}(t)^\frown\mathbf{plc}(t)^\frown\mathbf{ss}(l)^\frown\mathbf{se}\|}, & t \in L_T \\ d_0 \cdot \frac{\mathfrak{V}(t)^\frown\mathbf{plc}(t)^\frown\mathbf{0}^\frown\mathbf{se}}{\|\mathfrak{V}(t)^\frown\mathbf{plc}(t)^\frown\mathbf{0}^\frown\mathbf{se}\|}, & t \in T \end{cases}$$

$$\Theta_0'(t, \mathfrak{V}, \mathfrak{W}) = \{p \mid \|p - c_t\| < r_0\} \tag{7.7}$$

Geometric Connectionist Machine uses the *Abbeschriftung* operation abb to transform a labeled tree into an unlabeled tree $\mathsf{abb}(\mathfrak{L})$, and Θ_0 as the initialization operator, then re-uses $\mathbf{GCM}_0$ to spatialize labeled tree structures. That is,

$$\mathbf{GCM}_1(\mathfrak{L}, \mathfrak{V}) = \mathbf{GCM}_0(\mathsf{abb}(\mathfrak{L}), \mathfrak{V}) \tag{7.8}$$

7.3 The Setting of Experiments

7.3.1 Datasets

Knowledge Graphs for Triple Classification are normally generated from WordNet (1995) and Freebase (2008). WordNet is a large English lexical database, whose entities are called *synset*, each represents a distinct word sense. Freebase is a large Knowledge Graph about world facts. Following the evaluation strategies (Bordes et al. 2013; Socher et al. 2013; Wang and Li 2016; Ji et al. 2015), we used WN11, WN18, FB13 to generate datasets for classifying Triples of 1-to-N relations. We manually analyzed all the relations in the three datasets, and rename them in term of the containment relation of $\mathscr{N}$-Balls as follows: (1) there are totally 26 1-to-N relations in the three datasets, with id from 0 to 25. If r is the ith non-symmetric relation, triple (h, r, t) is named as $contain_i(h, t)$ in $\mathscr{N}$-Ball representation, the prefix *tr_* is added to $contain_i$, if r is transitive; (2) if r^{-1} is the inverse of r, triple (h, r^{-1}, t) will be transformed to (t, r, h). The whole relations used in $\mathscr{N}$-Ball representation are listed in Table 7.1.

Table 7.1 Mapping WN18/WN11/FB13 relations to $\mathcal{N}$-Ball relations

Id	WN18/WN11/FB13 relation	$\mathcal{N}$-Ball relation
0	A B _member_of_domain_topic	$contain_0$(A, B)
	A _domain_topic B	
1	A B _member_meronym	$tr\_contain_1$(A, B)
2	A B _member_of_domain_region	$contain_2$(A, B)
	A _domain_region B	
3	A B _hypernym	$tr\_contain_3$(B, A)
	A _type_of B	
4	A B _member_holonym	$tr\_contain_1$(B, A)
5	A B _instance_hypernym	$contain_5$(B, A)
	A _subordinate_instance_of B	
6	A B _member_of_domain_usage	$contain_6$(A, B)
7	A B _synset_domain_topic_of	$contain_0$(B, A)
	A _synset_domain_topic B	
8	A B _hyponym	$tr\_contain_3$(A, B)
9	A B _instance_hyponym	$contain_5$(A, B)
	A _has_instance B	
10	A B _synset_domain_usage_of	$contain_6$(B, A)
11	A B _has_part	$tr\_contain_{11}$(A, B)
12	A B _part_of	$tr\_contain_{11}$(B, A)
13	A B _synset_domain_region_of	$contain_2$(B, A)
14	A gender B	$contain_{14}$(B, A)
15	A nationality B	$contain_{15}$(B, A)
16	A profession B	$contain_{16}$(B, A)
17	A place_of_death B	$contain_{17}$(B, A)
18	A place_of_birth B	$contain_{18}$(B, A)
19	A location B	$contain_{19}$(B, A)
20	A institution B	$contain_{20}$(B, A)
21	A cause_of_death B	$contain_{21}$(B, A)
22	A religion B	$contain_{22}$(B, A)
23	A parents B	$contain_{23}$(B, A)
24	A children B	$contain_{24}$(A, B)
25	A ethnicity B	$contain_{25}$(B, A)

We transform Triples in datasets into $\mathcal{N}$-Ball representations, and remove duplicated Triples. WN11 dataset has 11 relations, 38696 entities, 112581 training Triples, 2609 valid Triples, 21088 testing Triples. In machine learning, valid dataset is used separately to adjust some hyper-parameters Bishop (2006), Goodfellow et al. (2016). Our Triple classification model does not have hyper-parameters, which needs to be tuned by sample data, so we integrate true Triples in the valid dataset into the training dataset.

Table 7.2 Datasets extracted from WN11, FB13, WN18

Dataset	#$\mathscr{R}$	#$\mathscr{E}$	#train Triple	#test Triple
WN11 $\mathscr{N}$-Ball	5	38,696	94,472	14,587
FB13 $\mathscr{N}$-Ball	13	75,043	30,6747	45,897
WN18 $\mathscr{N}$-Ball	7	40,943	60,491	32,264

- After transformation and reduction, WN11 $\mathscr{N}$-Ball dataset has 5 relations, totaling 94472 training Triples (91888 from WN11 training Triples, 2584 from WN11 valid Triples), and 20495 testing Triples.
- FB13 dataset is transformed the same way.
- After transformation and reduction, WN18 $\mathscr{N}$-Ball dataset only remains 135 testing Triples. We increased to 19387 testing Triples with true value, and the same number of Triples with false values, following the same setting used for WN11.

To predict truth-value of (h, r, x), our model must have at least one tail t of (h, r, t) in the training set. We remove all Triples in the testing set which do not have sample tails in the training set. Final dataset sizes are listed in Table 7.2.

7.3.2 The Design of Experiment and the Evaluation Protocol

In the transformation-based approach, a threshold δ_r is defined to evaluate a score function f: if a transformation score f of a Triple is below δ_r, this Triple will be predicted as true. We simulate this method to fit the $\mathscr{N}$-Ball setting as follows: To judge whether x is the tail of h such that (h, r, x), we initialize the $\mathscr{N}$-Ball of x, $\mathbb{B}(\mathbf{O}_x, r_x)$, as the same way as initializing other $\mathscr{N}$-Balls, and transform $\mathbb{B}(\mathbf{O}_x, rd_x)$ following the same sequence and transformations as the $\mathscr{N}$-Ball construction process of any true tail t of h, satisfying (h, r, t). If $\mathbb{B}(\mathbf{O}_x, rd_x)$ is located in the r-subspace of the $\mathscr{N}$-Ball of h, (h, r, x) will be judged as truth, otherwise false. We update the final radius rd_{h_r} with a ratio γ, to see performances of the prediction result. The Triple classification model $\mathscr{M}(h, r)$ is constructed as illustrated in Algorithm 2. We used three different entity-embeddings: TransE Bordes et al. (2013), TEKE_E and TEKE_H Wang and Li (2016) as pre-trained embeddings.

Algorithm 2: Whether the triple (h, r, x) holds in knowledge-graph KG

```
input : head h, relation r, entity x, knowledge-graph KG, ratio γ, entity-embeddings EV
output: True, if 𝒩-Ball of x is contained by 𝒩-Ball of h; False, otherwise

// get all known tails of h related with r in KG
tails = get_all_known_tails(h, r, KG);
if number_of(tails) > 0 then
    tChains = get_fine_grained_types(h, KG); // Construct the mushroom
        model, return (1) the 𝒩-Ball of h, (2) the
        initialization parameter, the sequence of geometric
        transformations applied for tails
    ballH, tranHis =nballs(h, r, tails, tChains, EV)
    // multiply γ to the radius of ballH
    ballH =enlarge_radius(ballH, γ)
    // construct 𝒩-Ball of x with tranHis
    ballX =construct_nball(x, tranHis, EV)
    if contained_by(ballX, ballH) then
        return True;
    else
        return False;
    end
else
    return False;
end
```

7.4 Experiment 6: Triple Classification for FB13 Dataset

Triples in FB13 dataset do not have hypernym relations. As a result, the length of the fine-grained typing chain is zero. This degrades the model into a trivial case: Tripe (h, r, x) holds, if and only if x is inside the r-subspace of h's $\mathcal{N}$-Ball. We draw the minimum r-subspace containing all known true tails. Using TEKE_E vectors, the accuracy reaches a maximum of 78%, which is almost the same as 77.4% reported in Wang and Li (2016). Though our datasets are not exactly the same as the benchmark datasets, this result is consistent with existing results in the literature. FB13 $\mathcal{N}$-Ball datasets and pre-trained entity-embeddings are free for download.[2]

7.5 Experiment 7: Triple Classification for WN11 Dataset

Triples in WN11 dataset have hypernym relations. Heights of type chains vary from 1 to 12. Precisions, recalls, and accuracies are illustrated in Figs. 7.4, 7.5, and 7.6. We expand the range of γ from 1.0 to 2.3, to see whether smaller γ can contribute to precision. Experiment results support that smaller γ greatly improves precisions,

[2] https://figshare.com/articles/FB13nball_zip/7294295.

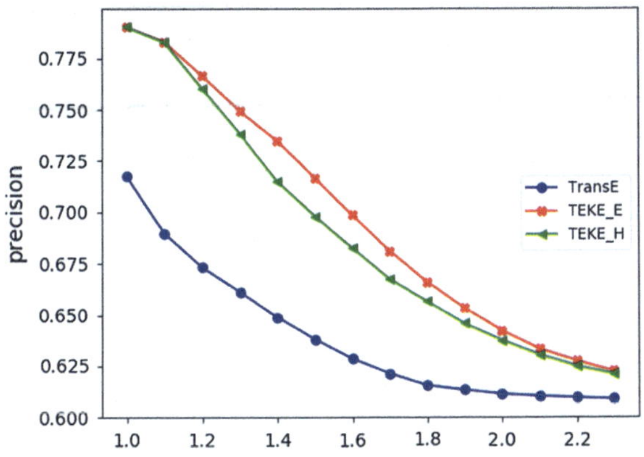

Fig. 7.4 Precision of Triple classification using WN11 $\mathcal{N}$-Ball dataset. When γ increases, the precision will drop

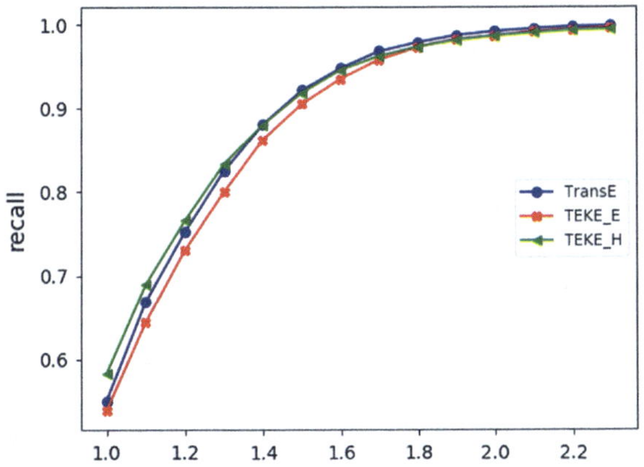

Fig. 7.5 Recall of Triple classification using WN11 $\mathcal{N}$-Ball dataset. When γ increases, the recall will increase

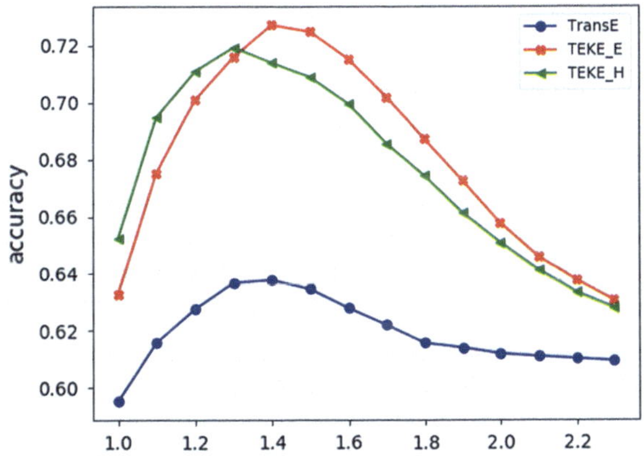

Fig. 7.6 Accuracy of Triple Classification using WN11 $\mathcal{N}$-Ball with different γ values

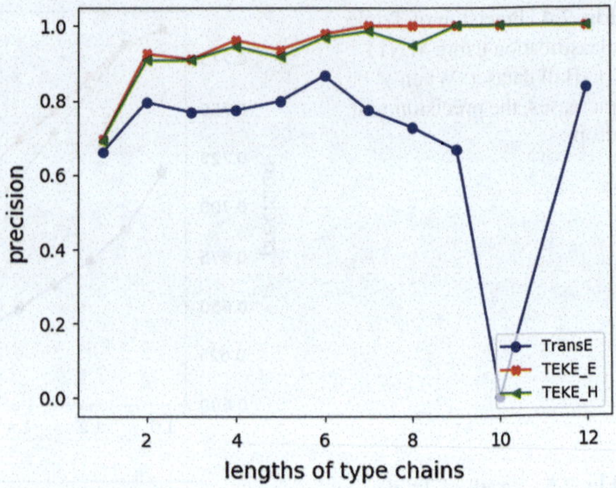

Fig. 7.7 Precision versus length of type chains in WN11 $\mathcal{N}$-Ball Dataset, when $\gamma = 1$

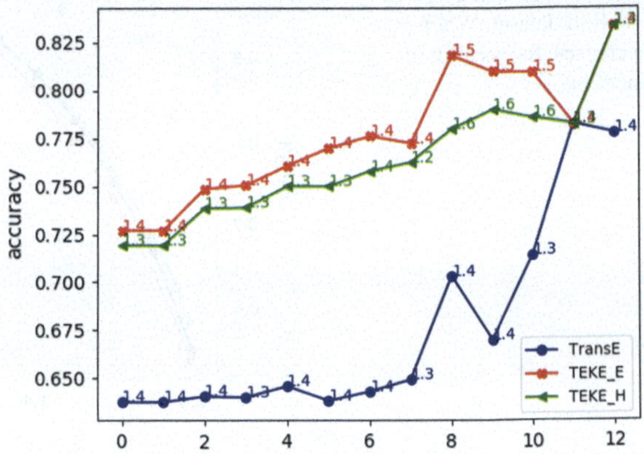

Fig. 7.8 Accuracy versus different lengths of type chains in WN11 $\mathcal{N}$-Ball Dataset; when lengths increases, the accuracy will have the strong tendency to increase; numbers in the plots represent the γ value with which the accuracy reaches maximum

and severely weakens recalls. The maximum value of 73% in accuracy is reached by using TEKE_E vectors when $\gamma = 1.4$, a bit less than 75.9% reported by Wang and Li (2016).

We have analyzed the contribution of lengths of type chains to precision. Choose $\gamma = 1.0$, Fig. 7.7 shows that lengths of type chains can contribute greatly to precision by using TEKE_E or TEKE_H entity-embeddings: If lengths are greater than 2, precisions will be greater than 90%, and will have a strong tendency to increase to 100% using both TEKE entity-embeddings. Such performance is not clear by TransE entity-embeddings, which suffers from a sudden drop with the length of type chains at 10. After examining the datasets, we find that there are only 19 testing records with type-chain length of 10, and that none of them is correctly predicted using TransE embeddings.

We have analyzed the relation between accuracies and lengths of type chains. Figure 7.8 shows the maximum accuracy with regard to Triples whose lengths of type chains are longer than N. For example, if we choose $\gamma = 1.5$, the accuracy

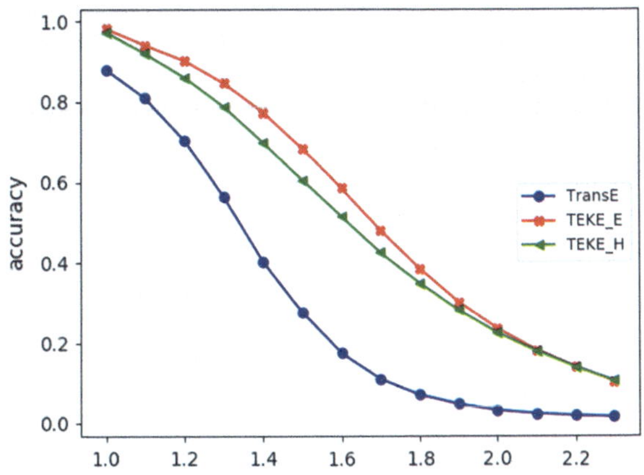

Fig. 7.9 With the unbalanced WN18 $\mathcal{N}$-Ball testing dataset, the accuracy reaches 98% using TEKE_E pre-trained entity-embeddings, when $\gamma = 1$

using TEKE_E embeddings will reach 81.8% for Triples whose type chains are longer than 8. When type chains are longer than 4, the accuracy of our model using TEKE_E embeddings will significantly outperform the results reported in Wang and Li (2016).

The performance by using TransE is not only lower than but also not as stable as that by using TEKE_E or TEKE_H entity-embeddings. The reason is that TEKE_E and TEKE_H entity-embeddings are jointly trained by knowledge-graph and corpus information, while TransE does not consider corpus information. WN11$\mathcal{N}$-Ball datasets and pre-trained entity-embeddings are free for public access.[3]

7.6 Experiment 8: Triple Classification for WN18 Dataset

Triples in WN18$\mathcal{N}$-Ball dataset have richer hypernym relations than Triples in WN11$\mathcal{N}$-Ball. For example, the lengths of type chains are longer than those of WN11$\mathcal{N}$-Ball: 57.1% type chains in WN18$\mathcal{N}$-Ball are longer than 5; the maximum length is 18. However, it turns out that only 72 tails of *true* testing Triples have pre-trained entity-embeddings, while 6574 tails of *false* testing Triples have. With this unbalanced testing dataset, our predicting model produces surprisingly good accuracy. Accuracy reaches 98% using TEKE_E pre-trained entity-embeddings, when $\gamma = 1$, as illustrated in Fig. 7.9. Lengths of type chains contribute to accuracy, see Fig. 7.10. TransE embeddings also delivered great results, though the performance remains less stable than those of TEKE_E and TEKE_H embeddings. WN18$\mathcal{N}$-Ball datasets and pre-trained entity-embeddings are free for download.[4]

[3] https://figshare.com/articles/WN11nball/7294307.

[4] https://figshare.com/articles/WN18nball/7294316.

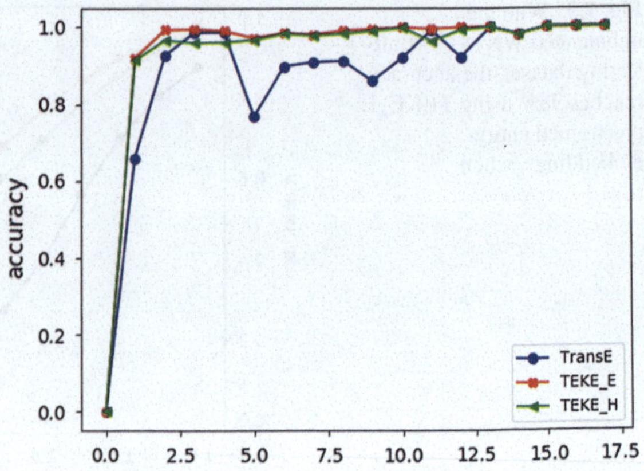

Fig. 7.10 With the unbalanced WN18 $\mathcal{N}$-Ball testing dataset, the accuracy increases along with the length of type-chains, when $\gamma = 1$

7.7 Summary

Triple Classification for 1-to-N relations is a tough task in Knowledge Graph reasoning. Symbolic approach is awkward, as it cannot deal with unknown symbols. The remedy is the learning representation approach that can do approximation by utilizing similarity relations among vector embeddings. There are some region-based approaches to embedding Knowledge Graphs. He et al. (2015) embed entities using Gaussian distributions—high-dimensional regions of probability, whose mean of the distribution is the center point, and whose covariance represents the uncertain boundary of a region. Nickel and Kiela (2017) use an unsupervised approach to embed tree structures into Poincaré balls. Xiao et al. (2016) represent Triple (h, r, t) as a manifold $\mathsf{M}(\mathbf{h}, \mathbf{r}, \mathbf{t}) = D_r^2$, that is, given h and r, the tail t shall be located in the manifold. In all of these works, structural embeddings have not been targeted at zero-energy cost, in part because the *limitivism* philosophy (see Sect. 2.3) believes it not possible. In this chapter, we present the novel Geometric Connectionist Machine approach to precisely (zero-energy cost) spatializing labeled tree structure onto vector embeddings. The Triple Classification problem is interpreted as a Way-finding task. Experiment results show that both symbolic structure and entity embeddings contribute to the performance.

References

Auer, S., Bizer, C., Lehmann, J., Kobilarov, G., Cyganiak, R., & Ives, Z. (2007). DBpedia: A nucleus for a web of open data. In K. Aberer, K.-S. Choi, N. Noy, D. Allemang, K.-I. Lee, L. Nixon, J. Golbeck, P. Mika, D. Maynard, R. Mizoguchi, G. Schreiber & P. Cudré-Mauroux (Eds.) *ISWC07*. Springer.

Bishop, C. M. (2006). *Pattern recognition and machine learning*. Secaucus, NJ, USA: Springer.

Bollacker, K., Evans, C., Paritosh, P., Sturge, T., & Taylor, J. (2008). Freebase: A collaboratively created graph database for structuring human knowledge. In *SIGMOD '08* (pp. 1247–1250) New York, NY, USA: ACM.

Bordes, A., Usunier, N., Garcia-Duran, A., Weston, J., & Yakhnenko, O. (2013). Translating embeddings for modelling multi-relational data. In C. J. C. Burges, L. Bottou, M. Welling, Z. Ghahramani, & K. Q. Weinberger (Eds.) *Advances in Neural Information Processing Systems, 26*, 2787–2795. Curran Associates, Inc.

Faruqui, M., Dodge, J., Jauhar, S. K., Dyer, C., Hovy, E., & Smith, N. A. (2015). Retrofitting word vectors to semantic lexicons. In *Proceedings of the 2015 Conference of the North American Chapter of the Association for Computational Linguistics: Human Language Technologies* (pp. 1606–1615). ACL.

Goodfellow, I., Bengio, Y., & Courville, A. (2016). *Deep learning*. The MIT Press.

Han, X., Liu, Z., & Sun, M. (2016). Joint representation learning of text and knowledge for knowledge graph completion. *CoRR*. abs/1611.04125.

He, S., Liu, K., Ji, G., & Zhao, J. (2015). Learning to represent knowledge graphs with gaussian embedding. In *CIKM'15* (pp. 623–632). New York, USA: ACM.

Ji, G., He, S., Xu, L., Liu, K., & Zhao, J. (2015). Knowledge graph embedding via dynamic mapping matrix. In *ACL'2015* (pp. 687–696). Beijing: ACL.

Manning, C. D., Raghavan, P., & Schütze, H. (2008b). *Introduction to information retrieval*. New York, NY, USA: Cambridge University Press.

Miller, G. A. (1995). Wordnet: A lexical database for english. *Communications of the ACM, 38*(11), 39–41.

Nickel, M., & Kiela, D. (2017). Poincaré embeddings for learning hierarchical representations. In I. Guyon, U. V. Luxburg, S. Bengio, H. Wallach, R. Fergus, S. Vishwanathan & R. Garnett (Eds.). *Advances in Neural Information Processing Systems* (Vol. 30, pp. 6338–6347). Curran Associates, Inc.

Socher, R., Chen, D., Manning, C. D., & Ng, A. (2013). Reasoning with neural tensor networks for knowledge base completion. In C. J. C. Burges, L. Bottou, M. Welling, Z. Ghahramani, & K. Q. Weinberger (Eds.) *Advances in Neural Information Processing Systems, 26*, 926–934. Curran Associates, Inc.

Speer, R., Chin, J., & Havasi, C. (2017). Conceptnet 5.5: An open multilingual graph of general knowledge. In *Proceedings of the Thirty-First AAAI Conference on Artificial Intelligence, February 4–9, 2017, San Francisco, CA, USA* (pp. 4444–4451).

Suchanek, F. M., Kasneci, G., & Weikum, G. (2007). Yago: A core of semantic knowledge. In *WWW '07* (pp 697–706). New York, NY, USA: ACM.

Wang, Z., & Li, J. (2016). Text-enhanced representation learning for knowledge graph. *IJCAI*, 1293–1299.

Wang, Z., Zhang, Feng, J., & Chen, Z. (2014a). Knowledge graph and text jointly embedding. In *EMNLP*, 1591–1601.

Xiao, H., Huang, M., & Zhu, X. (2016). From one point to a manifold: Knowledge graph embedding for precise link prediction. *IJCAI*, 1315–1321.

Xie, R., Liu, Z., Jia, J., Luan, H., & Sun, M. (2016). Representation learning of knowledge graphs with entity descriptions. *AAAI*, 2659–2665.

Zhang, D., Yuan, B., Wang, D., & Liu, R. (2015). Joint semantic relevance learning with text data and graph knowledge. *ACL-IJCNLP*, 32–40.

Zhong, H., Zhang, J., Wang, Z., Wan, H., & Chen, Z. (2015). Aligning knowledge and text embeddings by entity descriptions. *EMNLP*, 267–272.

Bollacker, K., Evans, C., Paritosh, P., Sturge, T., & Taylor, J. (2008). Freebase: A collaboratively [illegible] graph database for structuring human knowledge. In *SIGMOD* (pp. 1247–1250). New York, NY: ACM.

Bordes, A., Usunier, N., Garcia-Duran, A., Weston, J., & Yakhnenko, O. (2013). Translating embeddings for modeling multi-relational data. In C. J. C. Burges, L. Bottou, M. Welling, Z. Ghahramani, & K. Q. Weinberger (Eds.), *Advances in Neural Information Processing Systems 26* [illegible]. Curran Associates, Inc.

[illegible] (2018). [illegible] In *Proceedings of the* [illegible] *Conference of the North American Chapter of the Association for Computational Linguistics: Human Language Technologies* [illegible]

Goodfellow, I., Bengio, Y., & Courville, A. (2016). *Deep learning*. The MIT Press.

[illegible] Knowledge graph embedding [illegible]

[illegible] Knowledge graph embedding via dynamic mapping matrix. In *ACL* [illegible]

[illegible]

[illegible]

[illegible]

[illegible]

Chapter 8
Resolving the Symbol-Subsymbol Debates

The symbol spatialization process creates a new *geometric* layer between the connectionist layer and the symbolic layer. Connectionist networks produce vectors; Geometric layers promote these vectors into $\mathscr{N}$-Balls in higher dimensions; Symbolic structures are precisely encoded as spatial relations among these balls, as illustrated in Fig. 8.1. The existence of this new geometric layer resolves a variety of debates between connectionism and symbolicism.

8.1 Philosophies Behind the Symbol-Subsymbol Relation

As $\mathscr{N}$-Ball embeddings can only be created in higher dimensional space, we refute *eliminativism* that connectionist approach can achieve all symbolic approach can do, and refute *implementationalism* that neural-network is the "hardware" of symbolic system, also refute *revisionism* that a symbolic account can be generated by neural-networks. We favor both *limitivism* that accurate connectionist account can approximate good symbolic descriptions within certain limit, and *hybridism* that a *patchwork* can be created for the gap between symbolic and neural-network components. Here, Geometric Connectionist Machines are such geometric *patchworks* that are precisely located between symbolic and connectionist components.

8.2 Resolving a Variety of Symbol-Subsymbol Debates

$\mathbf{GCM}_0$ and $\mathbf{GCM}_1$ construct $\mathscr{N}$-Ball configurations of symbolic tree structures in a space whose dimension is higher than the dimension of vectors produced by connectionist networks. The $\mathscr{N}$-Ball configuration creates a continuum between connectionist networks and symbolic structures.

T. Dong, *A Geometric Approach to the Unification of Symbolic Structures and Neural Networks*, Studies in Computational Intelligence 910,
https://doi.org/10.1007/978-3-030-56275-5_8

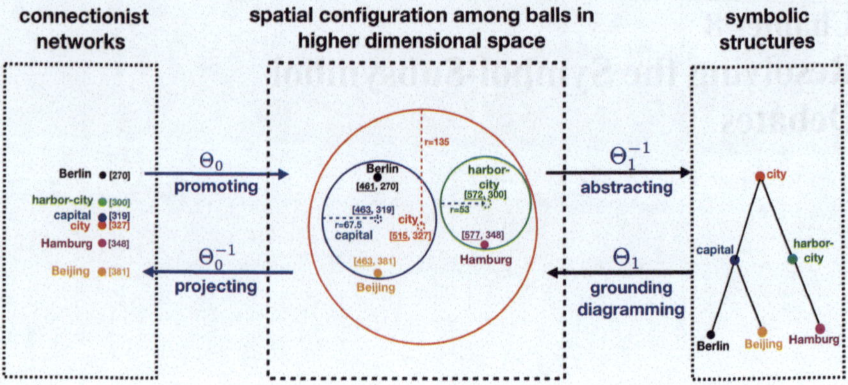

Fig. 8.1 Symbol spatialization reaches zero energy cost through geometric transformations

8.2.1 *The Necessity of Precisely Imposing External Knowledge onto Connectionist Networks*

Creating $\mathbf{GCM}_0$ and $\mathbf{GCM}_1$ is a concrete case study for the debate on the relation between Deep Learning and structure, shows the way to precisely impose tree structures onto vector embeddings, and favors the opinion that *structure is necessary good* (see Sect. 1.2).

An important research topics in AI and Cognitive Science is to automatically discover appropriate knowledge from data, (e.g., Kemp and Tenenbaum 2008; Grosse et al. 2012; Lu et al. 2019; Li et al. 2019). We do not believe it possible to learn all structures from the environment without imposing any external knowledge (one argument in the learning versus structure debate, see Sect. 1.2). That *stealing is not allowed* being set up as a law is due to the fact that *stealing* behaviors exist in the society. However, even *stealing* behaviors does not exist, it still holds that *stealing is not allowed* (see Sect. 1.1.1). The truth of laws is independent of data acquired from reality. Only fed with data of stealing behaviors, connectionist networks would be more likely to mimic stealing behavior, rather than to be enlightened that stealing is not allowed. Excluding all stealing data from the training set, connectionist networks may not learn the concept of *stealing* at all. Something external must be imposed onto the connectionist networks. This is not new to connectionists. In image recognition, they first precisely impose object names to each image in the training set. If they mistakenly imposed each cat image with the name ‘dog’, later their well-trained networks will recognize each cat image as ‘dog’. In unsupervised clustering task, they impose the number of clusters to the learning system ….

Without imposing any external knowledge, autonomous driving cars may have difficulty in learning whether they are in a right-hand traffic country, or in a left-hand traffic country. They even will not notice that there are two different traffic

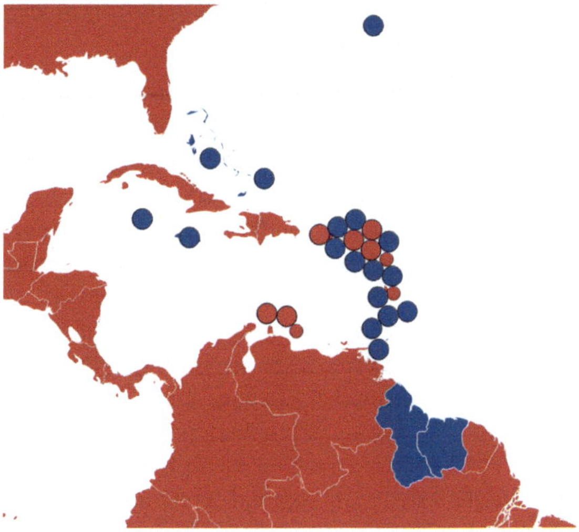

Fig. 8.2 Blue points represent British colonial islands with Left Hand Traffic System (LHTS); Red points represent French colonial islands with Right Hand Traffic System (RHTS). Without imposing any external knowledge, autonomous driving cars will not have concepts of different traffic systems. Picture is copied from Wikipedia

systems at all, especially when switching among right-hand traffic and left-hand traffic places, as illustrated in Fig. 8.2. Because concepts of *left* and *right* need an external reference framework (Levinson 1996), and country borders are subjectively determined by people in different nations, defended with their blood, even with their lives (Smith 2001). Which traffic rule on an island in Fig. 8.2 is determined not by the actual traffic scenario, rather by the fighting of colonists in the history—Left-hand traffic is imposed by British colonists, after they won the fighting.

The pre-condition of precisely imposing external knowledge onto connectionist network is that external knowledge can be precisely represented in the vector space. Our experiments show that the space for a precise representation of tree structured knowledge has higher dimension than the vector space produced by the connectionist network. This may lead to an *uncertainty principle* for knowledge discovery using connectionist approaches: *knowledge structures purely inferred from data through embedding approaches are uncertain, as the discovered knowledge structures in the embedding space is a projection of the true knowledge structures in an unknown higher embedding space*. This explains the reason that connectionist networks can only approximate symbolic reasoning within a limit (*limitivism*, see Sect. 2.3). This uncertainty principle suggests that precise imposition of external knowledge should be a must for real applications with high safety demand. Only after we precisely imposed traffic rules onto autonomous driving cars, can their driving behaviors be guaranteed and explained to be safe.

8.2.2 *Connectionism and Symbolicism are Compatible in $\mathcal{N}$-Ball Configurations*

Smolensky (1988b) claims that the incompatibility between connectionist and symbolic approaches is due to the different levels of analysis. Meanwhile, he somewhat contradictorily claims the possibility to obtain an accurate and complete account of cognitive process from both approaches. One way to resolve this contradictorily claim in the literature is to hold a co-evolution perspective that developments at one side will bring corresponding changes at other side, and in the end connectionist and symbolic models will be evolved into isomorphism.

Considering the discrete symbolic computation, (Rueckl 1988) doubted the existence of such isomorphism. Two particular examples are the membership assignment task and the application of production rules. Whether an entity is a member of a category? Whether the condition of a production rule is satisfied? From the classic symbolic perspective, both are discrete all-or-none tasks. One idea behind normal solutions is to *fuzzify* meanings of symbols (Zadeh 1965). It is a matter of degree whether a boy with the height of 1.70 m is a member of tall-boy class. We can map the truth degree with the pattern activation degree. Beside some theoretical limitations of the Fuzzy theory (Elkan 1993), the problem pointed out by Rueckl (1988) is that we need to know exactly which parts in a pattern are there, which are not there. Symbolic level descriptions, however, will not specify such information.

The $\mathcal{N}$-Ball configurations show a geometric way to Antony and Levine (1988), Dietrich and Fields (1988), Dyer (1988) how symbolic and connectionist approaches can be compatible, and refutes (Dietrich and Fields 1988)'s claim that the mapping between vectors at subsymbolic level and concepts at symbolic level should not be an approximation, for precise symbolic spatialization can be achieved in space of higher dimensions.

8.2.3 *$\mathcal{N}$-Ball Configuration as a Continuum between Connectionist Networks and Symbolic Structures*

Bechtel (1988) raised a challenging question as follows: If symbols have lower level implementations, what will be the lower level implementations of symbolic activities? He proposed an answer that *fixed, stable units that are stored in memory and retrieved into working memory where they are manipulated by rules*. It is neither clear how these units should be represented in working memory, nor clear how rules can be precisely implemented. The $\mathcal{N}$-Ball configuration constructed by GCM provides a clearer answer: As symbols are implemented by $\mathcal{N}$-Balls, and symbolic relations are implemented by spatial relations among $\mathcal{N}$-Balls, symbolic processing will be implemented by spatial operations on $\mathcal{N}$-Balls.

Nelson (1988) views the whole cognitive activity as follows: its upper level is a symbolic modeled in the traditional way, and is associated *in some way* to a fine-grained subsymbolic and parallel process modeled by connectionist networks. The GCMs is a concrete implementation to Nelson's *in some way*: the upper level symbolic model can be embodied as ball configurations in a higher dimensional space, the central points of balls are partially coincide with vectors produced by connectionist networks, if Nelson allows to add dimensions at all.

Prince and Pinker (1988) criticized that connectionist architectures are inadequate (Table 8.1), as connectionist models are limited to establish associations among subsymbols (Pinker and Prince 1988), and that Smolensky (1988b)'s analysis relies on *spurious theoretical* fusion as summarized in Table 8.2.

$\mathscr{N}$-Balls are easy to distinguish structural relations from similarity relations, types from tokens, meanwhile helping to keep individuals from blending (see Sect. 9.3). Subsymbolic vectors are well-preserved as a part of the center point of an $\mathscr{N}$-Ball. This shows a new way how connectionism could be integrated with symbolism and how subsymbolic issues are addressed within architectures of symbolic processing, as advocated by Prince and Pinker (1988).

Table 8.1 Connectionist architectures are inadequate as being lack of a number of important features (Prince and Pinker 1988)

Weakness of connectionist networks
Connectionist networks can not distinguish structural relations from similarity relations
Connectionist networks can not keep individuals from blending
Connectionist networks can not distinguish types from tokens
Connectionist networks can not selectively ignore similarity
Connectionist networks can not generate knowledge above and beyond trained associations

Table 8.2 Prince and Pinker (1988) claimed that a number of features that Smolensky (1988b) used to distinguish connectionism from symbolicism are not correct

Connectionism	Symbolicism	Why the mapping is spurious?
Sub-conceptual	Conceptual	• Theories for symbolic-processing *have no a priori commitment to* conceptual level • Sub-symbols in linguistics are handled by rules of an symbolic processing style
Parallel	Serial	• Theories for symbolic-processing *have no a priori commitment to strict seriality*
Context-sensitive	Context-free	• *Grammatical rules do not have consequences singly* (Pinker 1984)
Exact	Approximate	• Connectionism is exact at microscopic level, and approximate at macroscopic level • Symbolic processing is exact at macroscopic level, and approximate at microscopic level

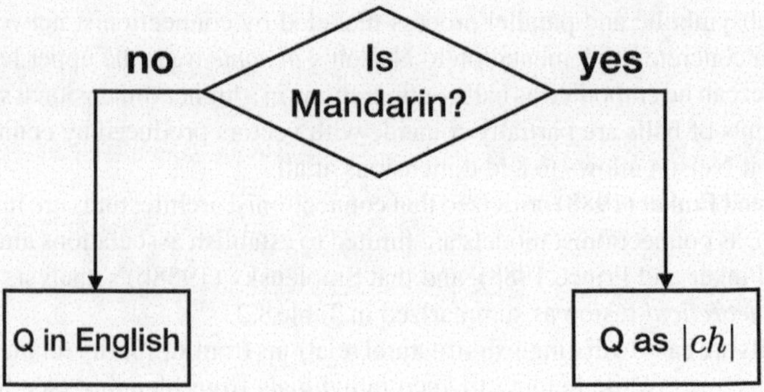

Fig. 8.3 An instruction that the letter Q shall be pronounced $|ch|$, if it appears in Mandarin

8.2.4 Resolving Epistemological Challenges

McCarthy (1988)'s challenge to connectionism is on epistemology issues, and questioned the *elaboration tolerance* of connectionist networks, which refers to the adaptability of the learning result to additional phenomena. That is, whether it can be elaborated with additional phenomena. One particular question raised by McCarthy was the pronunciation problem that appeared in the NETTalk system (Sejnowski and Rosenberg 1988). After sufficient training, NETTalk could translate English into speech. However, human English speakers can be further instructed that the letter Q shall be pronounced $|ch|$, if it appears in roman alphabet transcription of Mandarin, as illustrated in Fig. 8.3. How could this be realized by connectionists?

Smolensky (1988a) described his solution as follows: firstly, the rule on how to pronounce Q in Mandarin is *connectionist-implemented* rule interpreter as S-knowledge ('S' for Serial rule); in the intuitive processor are the P-knowledge ('P' for Pattern) of English pronunciation. When the system is reading English words, the computation is done in the intuitive processor. When the system encounters a Chinese word, the intuitive processor *fails to settle quickly on a pronunciation because of the non-English-like letter sequences*, so that the rule interpreter has a chance to trigger the Q pronounce in Mandarin. After some practice, the intuitive processor will update its weights to deal with the case of Q in Mandarin. In this way, S-knowledge is slowly and indirectly *compiled* into P-knowledge. Smolensky's solution would have unpredictable performances—the state of *failing to settle quickly* can represent other situations, such as, the system encounters Japanese, or unknown English words.

The problem can also be addressed using (Sun et al. 2005; Sun and Helie 2010)'s dual-process approach to simulating the interaction between explicit top-level symbolic representation and implicit bottom-level subsymbolic representation. Within the CLARION cognitive architecture (Sun 2016), we can describe the process as follows.[1]

[1] I am greatly indebted to Ron Sun for the personal communication.

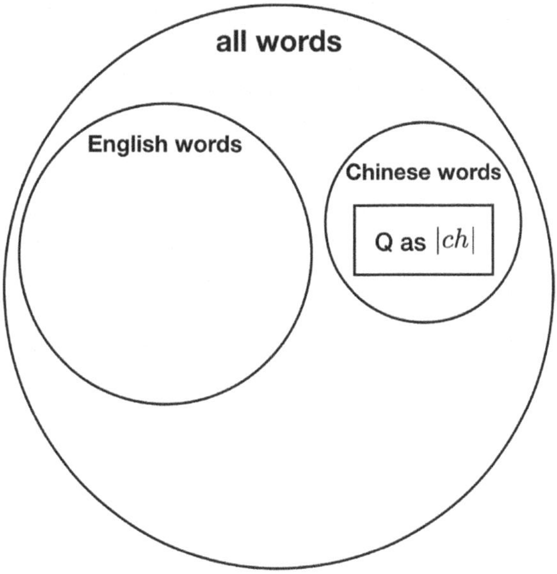

Fig. 8.4 An implementation using GCMs for an additional instruction

1. Let the current state x as *Q seems in roman alphabet transcription of Mandarin within an English context*.
2. Computing in the bottom level the "value" of two actions: $Q(x, a_e)$ and $Q(x, a_c)$, in which a_e is the action of pronouncing Q as an English letter, a_e is the action of pronouncing Q $|ch|$.
3. Find out possible actions at the top level using existing action rules. Here CLARION finds the rule *the letter Q shall be pronounced* $|ch|$, *if it appears in roman alphabet transcription of Mandarin*.
4. Choose an appropriate action a_e or a_c through integrating information of the top level and the bottom level. Here, CLARION will choose a_c (due to the combined information favoring a_c).
5. Perform the action a_c, and observe the next state y and the reinforcement.
6. Update the top level using the rule-extraction-refinement algorithm and update the bottom level using the reinforcement learning algorithm. Set current state x to *Q is in roman alphabet transcription of Mandarin*.

Different from Smolensky's solution and the Clarion model, GCM inserts the symbolic part of the representation directly into the subsymbolic part, and would be a clean and efficient solution. Multiple pronunciation of Q can be represented in the same way as multiple word-senses of a word are represented in Chap. 6. The additional instruction that the letter Q in Mandarin can be *connectionist-implemented* as new $\mathcal{N}$-Balls, in which nested relations encode the S-knowledge, and pronunciations within inner two $\mathcal{N}$-Balls encode P-knowledge, as illustrated in Fig. 8.4. Different from Smolensky, we encode S-knowledge as inclusion relations among regions, instead of activation patterns of vectors.

8.2.5 *Psychological Appeal for Continuous Representations*

The appealing feature of connectionism for psychologists is that connectionism captures events that either co-occur in space or time, or share similarities in meaning or in physics, while symbolism suffers from an unwanted *brittleness*: no guesses, no noisy stimulus (Smolensky 1988b). On the other hand, it is well agreed, even by Smolensky, that the fundamental function of cognition is inference (Dellarosa 1988). Inference has been precisely modeled in symbolic models. In most connectionist models, the fundamental process is on similarity, rather than on inference.

$\mathcal{N}$-Balls precisely encode symbolic tree structures in a continuous space, and well-preserve vectors produced by connectionist networks. Therefore, it has both appealing features above.

The existence of $\mathcal{N}$-Balls goes beyond the assumption that logical inference that can only be statistical approximated (Smolensky 1988b), pave the way to realize logical inference in continuous space, and also shakes the belief of Dellarosa (1988) that *connectionist models must demonstrate a greater degree of precision and accuracy in predicting and explaining the many nuances of human behavior than symbolic models currently do*—It would be easier for $\mathcal{N}$-Balls to take up this responsibility.

8.2.6 $\mathcal{N}$*-Ball Configuration Shapes Semantics*

Infants first make a categorical distinction between contact and non-contact (Carey 2009), and acquire three types of spatial relations in a specific order: topological relations, orientation relations, and distance relations (Piaget and Inhelder 1948). The topological relation between regions is the most fundamental relation in modeling spatial relations in the mind (Smith 1994). Taking the connection relation as primitive, we can construct distance comparison relations. Orientation relations can be constructed as a kind of distance comparison between one object and the sides of the other object, e.g. that *you are in front of the church* can be understood as *you are nearer to the main entrance of the church than to its other sides*. Distance relation can be further constructed as kind of distance comparison (see Sect. 2.1.1), e.g. that *an apple is one foot away from you* is understood as *an apple is nearer to the region of you extended by a one-foot sized object than to the region of you* (Dong 2008, 2012). We show that the disconnectedness relation and the part-of relation between open regions can be precisely constructed in higher dimension space in terms of the connection relation. This paves the way to solve the fundamental problem raised by Lakoff (1988) in *connectionist semantics*: How bounded regions and paths can be implemented in connectionist networks?

As psycholinguist, Lakoff favored continuous features, and agree with Pinker and Prince that *it is the connectionism, not the symbolic paradigm, that is the only game in town*. Lakoff introduced the term *connectionist semantics* that acquires ultimate meaning from the sensorimotor level and serves as meanings for the linguistic outputs. Lakoff neglected that symbolic structures can be precisely spatialized onto vector space, and intuitively classified *connectionist semantics* into the connectionist paradigm, due to its continuous feature. As connectionist networks only produce vectors, he had to admit that he did not know the implementation of these mechanisms in connectionist networks.

8.2.7 GCM for Instantly Updating Symbolic Knowledge

Smolensky (1988b) distinguished cultural knowledge as knowledge shared, codified, learned, and used by groups of people, and modeled by symbolic approaches. This raised the question on the formation of symbolic cultural knowledge from subsymbolic individual knowledge (Belew 1988).

When Bill Clinton won the President election in 1992, he became US President. This piece of cultural knowledge must be updated, and Smolensky (1988b) will have difficulty in using connectionist networks to simulate this update. As Belew (1988) commented *in fact, by current knowledge representation standards, connectionist nets are particularly simple: weighted digraphs*. The vector representing Bill Clinton, after adding the sentence *Bill Clinton is now US President* to the training corpus, will remain almost the same as before.

By utilizing Geometric Connectionist Machines, we update the $\mathcal{N}$-Ball embedding of Bill Clinton only by setting his PLC to US President and re-run the geometric construction process. After that, the new $\mathcal{N}$-Ball of Bill Clinton is inside of the $\mathcal{N}$-Ball of US President.

8.2.8 GCM Refuting the Metaphor of the Symbol-Subsymbol Relation to the Macro-Micro Relation in Physics

Smolensky (1988b) likens the relation between symbolic and subsymbolic approaches to the relation between macro and micro theories in physics, such as classical and quantum mechanics, classical and statistical thermodynamics. He claims that symbolic models describe macro-structural cognition and subsymbolic models describe the micro-structure of cognition and that symbolic accounts shall be *reduced* to subsymbolic ones. Smolensky (1988b) believes that the complete formal account of cognition lies at the sub-conceptual level and that subsymbolic accounts are more fundamental than symbolic ones.

Cleland (1988) pointed out that the reduction of macro-physical theories into micro-physical theories of physics is by no means trivial and that no one has been able to state laws to bridge them. Therefore, Smolensky (1988b)'s metaphor is not obvious. Cleland (1988) suggested that a relation of *reducibility between symbolic and subsymbolic in which conceptual is not autonomous from the sub-conceptual.*

Without a clear understanding of units and weights in the connectionist network, Cleland (1988) could not imagine how to do this.

Geometric Connectionist Machine is a bridge between connectionist networks and symbolic systems. As the dimension of $\mathscr{N}$-Balls constructed by Geometric Connectionist Machines are higher than that of vectors produced by connectionist network at sub-conceptual level, symbolic structures cannot be autonomously generated from the subsymbolic vectors.

8.2.9 GCM as a Content-Addressable Memory

Hunter (1988) severely criticized Smolensky (1988b)'s connectionism. For Hunter, Smolensky (1988b)'s connectionism is only feed-forward networks trained either by simulated annealing or by back-propagation of errors, ending up with a simulation of *content-addressable memory* that can date back to Luria (1966), in which the address can be figured out by the content. That is, from what we can know where. However, in Smolensky (1988b)'s connectionism, addresses can only be approximated by similarity without boundary. After training, similar contents shall be located in similar locations, but it is not a crisp region whose boundary can be determined by a similarity function. The Geometric Connectionist Machines are more like content-addressable memories. If we know the category of an entity, we will know in which ball this entity shall be located. GCM also provides a route in terms of a sequence of geometric transformations, to guide this entity to reach the target location.

8.2.10 GCM Saves Two Birds Through a Marriage

Woodfield (1988) considered the two-system hypothesis of the mind, one for symbolic, the other for subsymbolic, and listed three possible ways for the interaction between the symbolic system and the subsymbolic system.

- Division of labor. The symbolic system is operated by one part of the brain, while the subsymbolic system is operated by another part.
- Killing two birds with one stone. The two systems are operated by the same neural event that is capable of simultaneously manipulate conceptual-level symbols and perform subsymbolic operations.
- The "by" relation—The symbolic process emerges out of the subsymbolic.

Among the three possibilities, Woodfield (1988) favor the "by" relation, and proposed the terminology *connectionist machine* that *can, under certain conditions, simulate a von Neumann machine*. However, GCM refutes this "by" relation, because symbolic structures can only be spatialized in higher dimensional spaces, so cannot emerge out of the subsymbolic that exists in lower dimensional spaces. GCM also refutes the way of *killing two birds with one stone*, as GCM kills neither of the two.

Instead, GCM let the two birds marriage and form a ball structured family: the subsymbolic bird structures a part of the central point, the symbolic bird structures the rest part of the central point and the length of the central point vector and the radius. GCM turns the discrete symbolic bird into a continuous one (see Sect. 8.2.5).

8.3 Summary

There is a strong debate in the literature on the relation between connectionist model and symbolic model. Geometric Connectionist Machines serve as a step-stone between connectionist and symbolic models, and resolves almost all the debates.

References

Antony, L., & Levine, J. (1988). On the proper treatment of the connection between connectionism and symbolism. *Behavioral and Brain Sciences, 1*, 23–44.
Bechtel, W. (1988). Connection and interlevel relationism. *Behavioral and Brain Sciences, 1*, 24–25.
Belew, R. K. (1988). Two constructive themes. *Behavioral and Brain Sciences, 1*, 25–26.
Carey, S. (2009). *The origin of concepts*. Oxford University Press.
Cleland, C. E. (1988). Is Smolensky's treatment of connectionism on the level? *Behavioral and Brain Sciences, 1*, 27–28.
Dellarosa, D. (1988). The psychological appeal of connectionism. *Behavioral and Brain Sciences, 1*, 28–29.
Dietrich, E., & Fields, C. (1988). Some assumptions underlying Smolensky's treatment of connectionism. *Behavioral and Brain Sciences, 1*, 29–31.
Dong, T. (2008). A comment on RCC: From RCC to RCC^{++}. *Journal of Philosophical Logic, 37*(4), 319–352.
Dong, T. (2012). *Recognizing variable environments—The theory of cognitive prism*, Volume 388 of Studies in computational intelligence. Berlin, Heidelberg: Springer.
Dyer, M. G. (1988). The promise and problems of connectionism. *Behavioral and Brain Sciences, 1*, 32–33.
Elkan, C. (1993). The paradoxical success of fuzzy logic. *IEEE Expert*, 698–703.
Grosse, R. B., Salakhutdinov, R., Freeman, W. T., & Tenenbaum, J. B. (2012). Exploiting compositionality to explore a large space of model structures. In *Proceedings of the Twenty-Eighth Conference on Uncertainty in Artificial Intelligence, UAI'12* (pp. 306–315). Arlington, Virginia, USA: AUAI Press.
Hunter, L. E. (1988). Some memory, but no mind. *Behavioral and Brain Sciences, 1*, 37–38.
Kemp, C., & Tenenbaum, J. B. (2008). The discovery of structural form. *Proceedings of the National Academy of Sciences, 105*, 10687–10692.
Lakoff, G. (1988). Smolensky, semantics, and the sensorimotor system. *Behavioral and Brain Sciences, 1*, 39–40.
Levinson, S. (1996). Frames of reference and Molyneux's question: Cross-linguistic evidence. In P. Bloom, M. A. Peterson, L. Nadel, & M. F. Garrett (Eds.), *Space and language* (pp. 109–169). Cambridge: MIT Press.
Li, X., Vilnis, L., Zhang, D., Boratko, M., & McCallum, A. (2019). Smoothing the geometry of box embeddings. In *International Conference on Learning Representations (ICLR)*.

Lu, S., Mao, J., Tenenbaum, J., & Wu, J. (2019). Neurally-guided structure inference. In K. Chaudhuri & R. Salakhutdinov (Eds.), *Proceedings of the 36th International Conference on Machine Learning*, Volume 97 of Proceedings of Machine Learning Research (pp. 4144–4153). Long Beach, California, USA: PMLR.
Luria, A. R. (1966). *Higher cortical functions in man*. USA: Springer.
McCarthy, J. (1988). Epistemological challenges for connectionism. *Behavioral and Brain Sciences*, *1*, 44.
Nelson, R. J. (1988). Connections among connections. *Behavioral and Brain Sciences*, *1*, 45–46.
Piaget, J., & Inhelder, B. (1948). *La représentation de l'espace chez l'enfant*. Bibliothèque de Philosophie Contemporaine, Paris: PUF. English translation by F. J. Langdon and J. L. Lunzer in 1956.
Pinker, S. (1984). *Language learnability and language development*. Harvard University Press.
Pinker, S., & Prince, A. (1988). On language and connectionism: Analysis of a parallel distributed processing model of language acquisition. *Cognition*, *28*, 73–193.
Prince, A., & Pinker, S. (1988). Subsymbols aren't much good outside of a symbol-processing architecture. *Behavioral and Brain Sciences*, *1*, 46–47.
Rueckl, J. G. (1988). Making the connections. *Behavioral and Brain Sciences*, *1*, 50–51.
Sejnowski, T. J.,& Rosenberg, C. R. (1988). Neurocomputing: Foundations of research. In *NETtalk: A parallel network that learns to read aloud* (pp. 661–672). Cambridge, MA, USA: MIT Press.
Smith, B. (1994). Topological foundations of cognitive science. In C. Eschenbach, C. Habel, & B. Smith (Eds.), *Topological foundations of cognitive science*. Buffalo, NY: Workshop at the FISI-CS.
Smith, B. (2001). Fiat objects. *Topoi*, *20*(2), 131–148.
Smolensky, P. (1988a). Putting together connectionism—Again. *Behavioral and Brain Sciences*, *1*, 59–70.
Smolensky, P. (1988b). On the proper treatment of connectionism. *Behavioral and Brain Sciences*, *1*, 1–23.
Sun, R. (2016). Implicit and explicit processes: Their relation, interaction, and competition. In L. Macchi, M. Bagassi, & R. Viale (Eds.), *Cognitive unconscious and human rationality* (pp. 27–257). Cambridge, MA: MIT Press.
Sun, R., & Helie, S. (2010). Incubation, insight, and creative problem solving: A unified theory and a connectionist model. *Psychological Review*, *117*(3), 994–1024.
Sun, R., Slusarz, P., & Terry, C. (2005). The interaction of the explicit and the implicit in skill learning: A dual-process approach. *Psychological Review*, *112*(1), 159–192.
Woodfield, A., & Morton, A. (1988). The reality of the symbolic and subsymbolic systems. *Behavioral and Brain Sciences*, *1*, 58.
Zadeh, L. A. (1965). Fuzzy sets. *Informations and Control*, *8*, 338–353.

Chapter 9
Conclusions and Outlooks

What would be an interesting and tendentious claim is that there's no distinction between rule-following and rule-violating mentation at the cognitive or representational or symbolic level.
(Fodor and Pylyshyn 1988, p. 11)

Capable of classifying huge amount of texts, translating hundreds of languages, predicting the rise and fall of global markets, even driving unmanned automobiles, Deep Learning systems are the hope of the fifth industrial revolution. However, recent studies have found that Deep Learning systems can be easily manipulated, i.e., in image recognition and in natural language understanding. The nature of one system of the mind (System 1), which Deep Learning systems simulates, dictates that any given data would be put into a coherent story, even at the cost of logic. Another system of the mind (System 2) manages logical thinking by following rules and structures, which symbolic AI simulates. How to combine Deep Learning with symbolic structures remains an open debate. This debate is nothing new and dates back to the antagonism between Connectionism (neural-networks, now evolved into Deep Learning systems) and Symbolicism (symbolic representation and reasoning). The difficulty of this integration is that encoding symbolic relations cannot be achieved precisely by updating vector parameters using the back-propagation method, as described in Chap. 4. We illustrates a new geometric transformation method, GCM, to promote vectors onto higher dimensional $\mathcal{N}$-Balls. If we represent $\mathcal{N}$-Balls as vectorial forms, we can feed them into the next Deep Learning system, and have a new architecture, as illustrated in Fig. 9.1.

T. Dong, *A Geometric Approach to the Unification of Symbolic Structures and Neural Networks*, Studies in Computational Intelligence 910,
https://doi.org/10.1007/978-3-030-56275-5_9

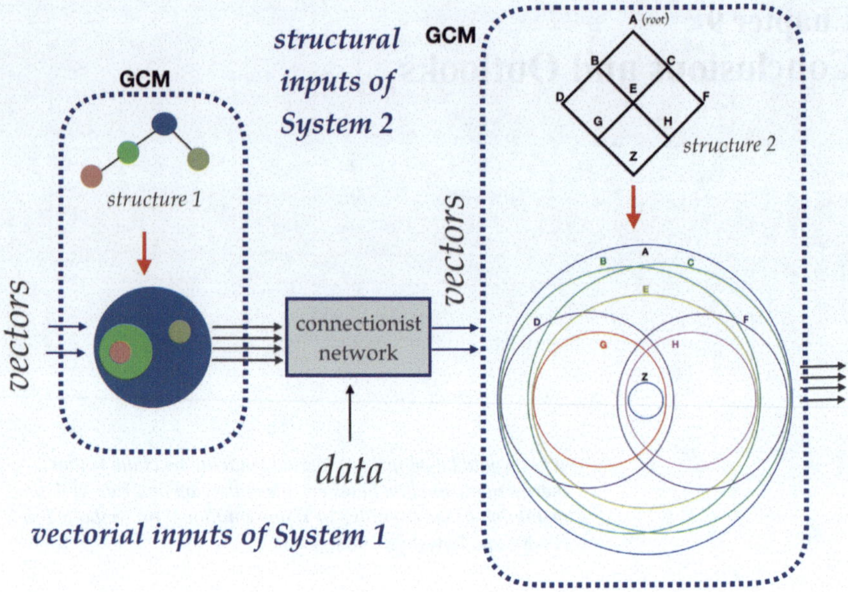

Fig. 9.1 Synergistic integration of two-system model of the mind through Geometric Connectionist Machine (GCM)

9.1 Structural Imposition onto Empty Vectors

One extreme case for $\mathcal{N}$-Ball embedding is to impose symbolic structures onto empty vector embeddings. For example, imposing tree structures onto empty vector embeddings will produce nested $\mathcal{N}$-Balls encoding child-parent relations and disconnected $\mathcal{N}$-Balls encoding sibling relations. This will end with diagram representation of the symbolic structure, similar to Venn diagram (Venn 1880), Euler diagram, and connect $\mathcal{N}$-Ball research with Diagrammatic Reasoning (DR) (Glasgow et al. 1995), even Cognitive Maps (Tolman 1948), Cognitive Collages (Tversky 1993; Tversky 2019, p. 82). In non-extreme cases (imposing symbolic structures onto non-empty vector embeddings), reasonings with DR and Cognitive Maps shall be enhanced by Deep Learning Systems.

9.2 Informed Machine Learning

Machine learning, especially Deep Learning, has made great success, and continues to shape the future of AI and our everyday lives, as predicted by Cremers et al. (1994). However, purely data-driven machine learning suffers from a number of issues mostly related with the insufficiency of input data, the robustness and security of learning results.

(a) Benin is a country near to the equator

(b) Pelican, a white bird

Fig. 9.2 "White as snow" is translated into "white as pelican" in the Natemba language

Hunter (1988) argued that connectionism is not a complete theory for learning. Learning through huge amount of data using back-propagation is inefficient, and only establishes a harmony between input and output. An important learning style in schools and universities is learning under instruction, or *top-down learning* (Sun 2016). It is easy to translate "white as snow" into German ("weiß wie Schnee"), into French ("blanc comme neige"), and many other languages. How shall we translate it into the Natemba language in which there is no word for snow? People who speak Natemba live in Benin, a country near to the equator, where the temperature is around 20 °C in the winter, so no snowing in the winter, as a consequence, no word for snow in the Natemba language. To describe something very white, people would say "white as pelican" (an example of *elaboration tolerance* in translation, see Sect. 8.2.4), as illustrated in Fig. 9.2. We feel this translation interesting and reasonable, after having been informed about the background knowledge. We may be surprised, if a machine learning system aligns "snow" with "pelican" with a low probability, either holding this alignment as false—"snow" should not be aligned with "pelican", or confirming the correctness, if we know that in the metaphor of the Natemba language, "snow" should be aligned with "pelican". Methods should be developed to precisely inform (or impose) external knowledge to pure data-driven machine learning systems.

Integrating knowledge into machine learning is not new in the literature and dates back to Towell and Shavlik (1994). A recent survey on this topic is conducted by von Rüden et al. (2019) who propose the term *Informed Machine Learning* to distinguish from the paradigm of pure data-driven machine learning. Bauckhage et al. (2019a) view Deep Learning as functional composition by writing as an L-layered connectionist network as follows.

$$\mathbf{y}(\mathbf{x}) = f_L(\ldots f_2(f_1(\mathbf{x})))$$

in which f_l is the function representing the computation of lth layer in the network. Let $\mathbf{z}_{l-1}$ and $\mathbf{z}_l$ be the input and the output of the lth layer, respectively, we should have

$$\mathbf{z}_l = \sigma_l(\mathbf{W}_l(\mathbf{z}_{l-1}))$$

where σ_l and $\mathbf{W}_l$ are the activation function and the matrix of connection weights, respectively.

If domain knowledge can also be represented by a set of real functions from $\mathbb{R}^m$ to $\mathbb{R}^n$, integrating external domain knowledge with connectionist networks will be seamlessly interpreted as functional composition, as illustrated by Bauckhage et al. (2019a) in a case study of solving the XOR classification task (Minsky and Papert 1988).

Domain knowledge normally appears in linguistic forms. Vectors from connectionist networks encode features of domain entities (Smolensky 1988a), domain knowledge shall structure their multi-arity relations. There is no reason to believe that each multi arity relation can be encoded within the space structured by unary feature vectors. Tree-structured domain knowledge can be precisely integrated with connectionist networks, and included in the framework of *Informed Machine Learning* by introducing four functions as follow.

- a function that extends dimension of a vector $f_{ext}: \mathbb{R}^m \to \mathbb{R}^{m+k}$
- a function that performs homothetic transformation $f_{homo} = \mathsf{H}$
- a function that performs shift transformation $f_{shift} = \mathsf{S}$
- a function that performs rotation transformation $f_{rota} = \mathsf{R}$.

To precisely integrate structured linguistic knowledge with connectionist network, We propose that *Informed Machine Learning* shall utilize region-based representation (*regionism*), whose philosophy dates back to de Laguna (1922), Whitehead (1929), and use novel optimization methods, instead of back-propagation. Linguistic terms, or concepts, can be learned and prototyped in terms of *minimum enclosing balls* using simple recurrent neural-networks (Bauckhage et al. 2019b).

9.3 A New Building Block for Connectionist Networks

The computational model of connectionist networks is inspired by observing the function of biological neuron, which contains soma, dendrites, and an axon. Messages are carried by *ions* that pass through neurons. The observability at least suggests that networks constructed by ions and neurons have physical sizes. It would be an unwritten chapter to promote the building block of connectionist network from vectors to regions. The simplest region is the ball in high-dimensional space, which only needs us to add one parameter of radius to the original vector. If we set all radii to the same extremely small positive real number $\epsilon > 0$, the representation meaning of this connectionist network will be the same as the connectionist network when all radii are set to 0.

This new building block is more expressive to represent recursive structures in semantic network, in the way solving the crosstalk problem—how shall we construct semantic networks to represent *yesterday John bought apples* and *last week John bought pears*, as shown in Fig. 9.3. If *John* refers to the same individual in the two sentence, we shall integrate their semantic structure together, as shown in Fig. 9.4. Now, it is not clear whether John bought apples yesterday or last week.

If connectionist networks produce balls, things will be different. John is a creature with a temporal span from the birth to the death, among which there exists two temporal parts: yesterday and last week. *Bought* is an event with a temporal duration. All these are SPAN ontologies (Grenon and Smith 2004), and represented by nested balls according to their temporal inclusion relations. In this representation framework, balls representing *apples* are located inside *yesterday* ball, while *pears* are located inside *last week* ball, as illustrated in Fig. 9.5.

Symbolic-manipulation, such as production rules, hierarchical tree, is a challenge task for connectionist networks. In the paradigm of connectionism, symbols, both

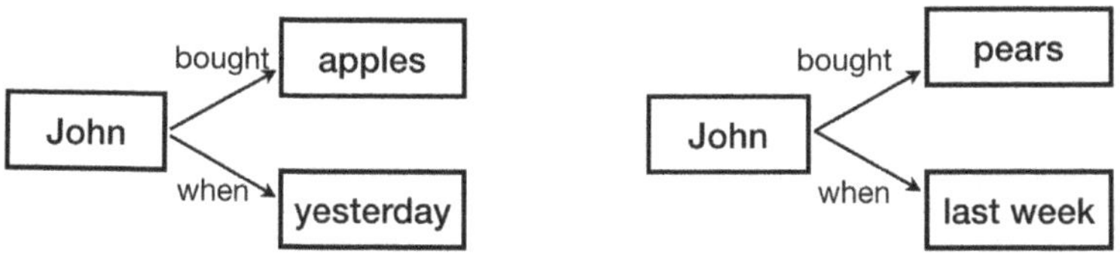

Fig. 9.3 Semantic structures of *yesterday John bought apples; last week John bought pears*, respectively

Fig. 9.4 The integrated semantic network

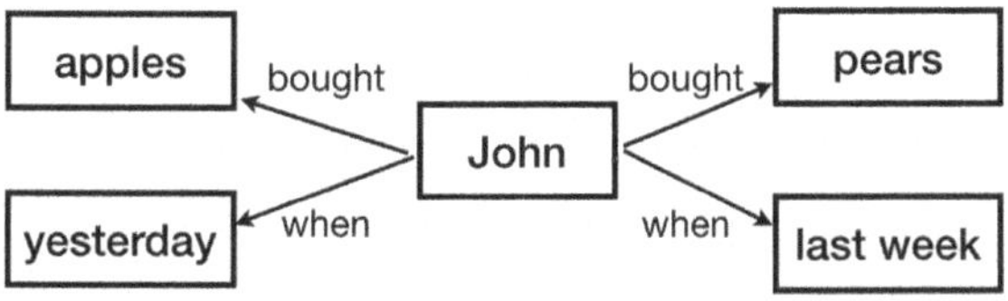

Fig. 9.5 An $\mathcal{N}$-Ball solution for the crosstalk problem

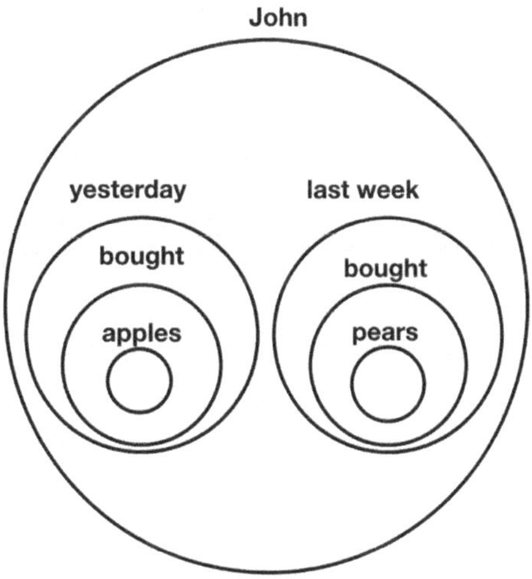

entities and relations, are represented by vectors. Marcus (2003) specifies symbolic-manipulation as three separate functions as follows.

- Representing abstract relationships between variables
- Representing recursive structures
- Distinguishing representations of individuals from representations of classes.

None of them can be easily handled by connectionist networks. In order to learn the *identity mapping* from domain **X** to range **Y** as listed in Table 9.1, people may develop a connectionist network with one hidden layer as shown in Fig. 9.6.

However, this network, as pointed by Marcus (2003), has no single sample in the training set to support that '1' as the right most digit in **X** should be mapped to '1' as the right most digit in **Y**.

A nice way to explain is to use two-system model of the mind: The fast model (System 1) checks each pair, with four digits as a whole. After 4 times of checking, System 1 makes a coherent story: each element in **X** is mapped to the same element

Table 9.1 Sample dataset for the identity mapping from **X** to **Y**. This example is copied from Marcus (2003, p. 37)

X	**Y**
1010	1010
0100	0100
1110	1110
0000	0000
1111	?

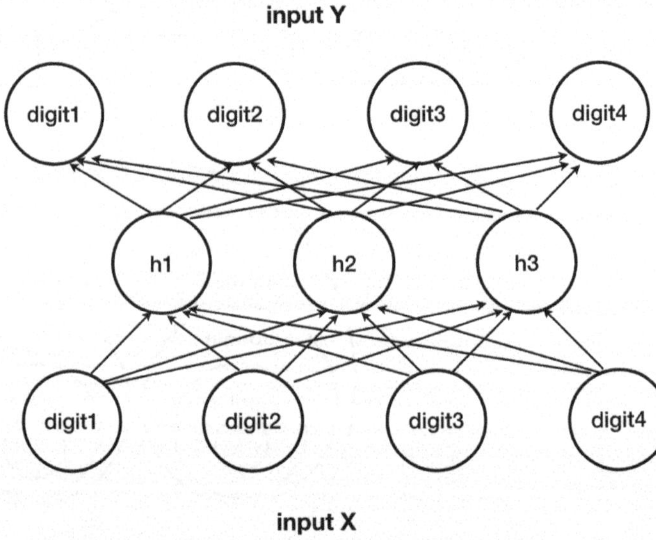

Fig. 9.6 A connectionist network to learn the identity relation. This diagram is copied from Marcus (2003, p. 46)

in **Y**. So, '1111' in **X** shall be mapped to '1111' in **Y**. The slow model (System 2) is skeptical, and conducts detailed checking for each pair, digit to digit. After 16 times checking, System 2 corrects the quick decision of System 1: the last digit in **X** should be mapped to '0' in **Y**.

The corresponding explanation using ball embedding is as follows: The fast model views members of **X** structure ball $\mathbb{B}_X$, members of **Y** structure ball $\mathbb{B}_Y$. The mapping from **X** to **Y** only needs one parameter, as illustrated in Fig. 9.7. Let $Y = f(X) = aX$, the back-propagation method will find out $a = 1$ using the four samples. '1111' in **X** is therefore mapped to '1111' in **Y**. In the slow thinking model, mappings are carried out at column level. The ball of the last column of **Y** contains a constant '0'. So, '1111' would be mapped to '1110'. In this way, there is no absolute truth criterion for the mapping. Some people may map to '1111', others map to '1110'. It is a window to explore the way of thinking (Tversky 1992, p. 131).

It is important for us to distinguish individual from category, or among individuals. Horse, as a category, has features such as four legs, two eyes, one tail. If we represent *horse 1* using features *X, Y, Z*, we will use the same features to represent its twin *horse 2*. This leads to an unintended representation that both *horse 1* and *horse 2* have the same feature vector. This is the well-known *two-horse problem* (Marcus 2003, p. 123). The same happens to train word (entity) embeddings, connectionist networks do not distinguish individuals from their categories. The training target is that words (entities) located in the similar context shall have similar vector embeddings. As words (entities) in the same category are often located in the same context, they have very similar vector embeddings. This loses the representation power to distinguish individuals from their categories. A promising solution is to represent categories as regions, e.g., balls: Individuals are located inside their category regions (see Sect. 6.5.3) (Lv et al. 2018).

9.4 Language Acquisition

In the past fifty years, linguists believed the existence of some abstract concepts that involve with a cognitive model of language, without discovering procedures

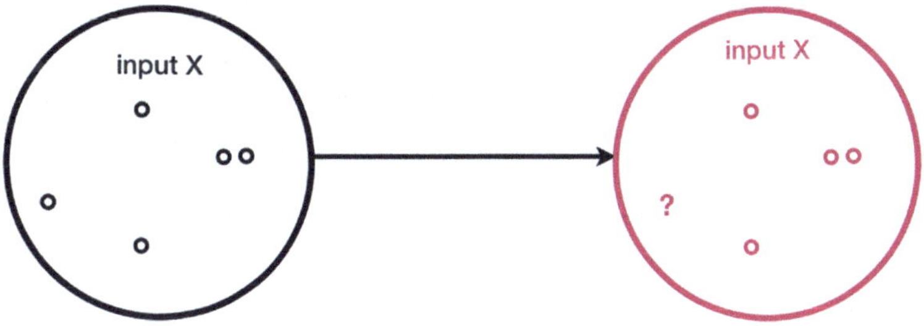

Fig. 9.7 An $\mathcal{N}$-Ball solution, in the slow thinking model, to predict the identity relation

for grammars. Freidin (1988) claims that it would be impossible that these abstract concepts could emerge from any kind of connectionist networks, for the jobs of connectionist networks are limited with statistical analysis of data (one more evidence for the necessity of precisely imposing external knowledge onto connectionist networks, Sect. 8.2.1). It is the abstract concepts embedded in the formulation of general grammatical principle that separate ungrammatical sentences from grammatical ones. Freidin (1988) illustrated this by using the bound anaphors (yourself, myself, herself, ...): there is a well-known general condition *Principle A of the Binding Theory* dictating that the antecedent of an anaphor shall be its syntactic subject. In the following sample sentences used by Freidin (1988), the effect of Principle A is to dictate Sentence 1 as ungrammatical, Sentence 2 as grammatical.

1. *Mary_i believes Bill to like herself_i.
2. Mary_i believes herself_i to like Bill.

Principle A is formulated on, or abstracted from, ill-formed linguistic structures, therefore, language learners are not capable of generating such principles that they follow. The point here is that Principle A is part of the innate cognitive structure that children have (Freidin 1988). Without these innate structures, connectionist models, after learned Sentence 2 as grammatical, cannot decide whether Sentence 1 is ungrammatical or having a novel grammatical structure. Using ball configurations,

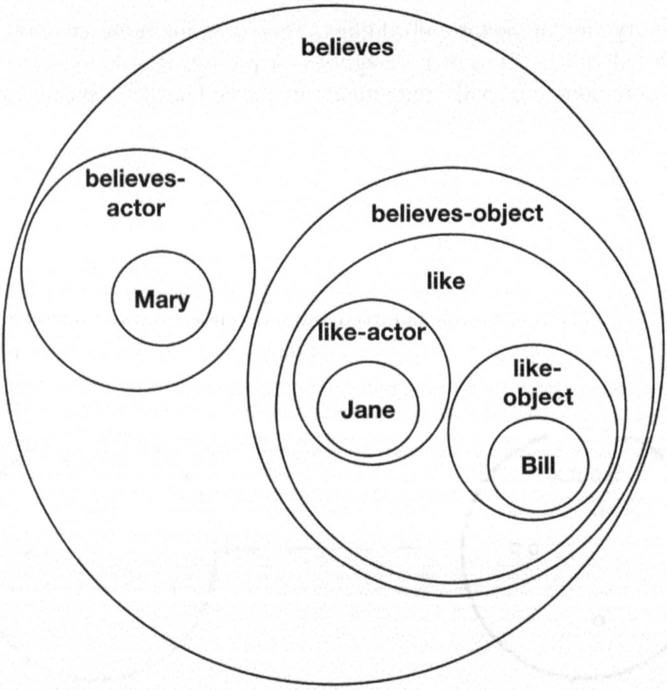

Fig. 9.8 The $\mathcal{N}$-Ball semantic representation of the sentence "Mary believes Jane to like Bill"

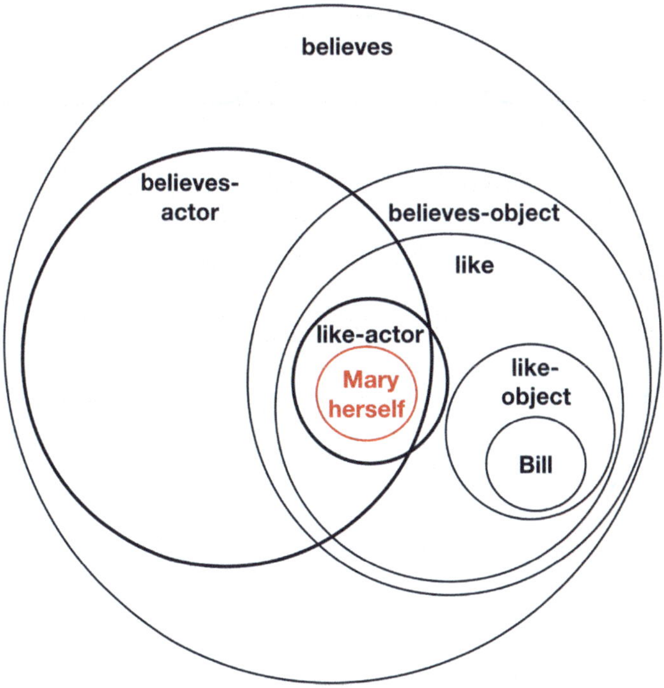

Fig. 9.9 The $\mathcal{N}$-Ball of believes-actor is enlarged to partially overlap with $\mathcal{N}$-Ball of like-actor, so that $\mathcal{N}$-Ball of Mary could coincide with $\mathcal{N}$-Ball of Jane. This ends with the denotation of herself

we may be able to explain Principle A in terms of certain rules of geometric transformations. Let us consider the sentence without anaphors:

Mary believes Jane to like Bill.

We design an $\mathcal{N}$-Ball configuration as the meaning of this sentence. Relative word order is captured by the distance relation among $\mathcal{N}$-Ball of words. For example, if 'Mary' is nearer to 'Jane' than to 'Bill' in the sentence, then $\mathcal{N}$-Ball of Mary will be nearer to $\mathcal{N}$-Ball of Jane than to $\mathcal{N}$-Ball of Bill, as illustrated in Fig. 9.8.

The $\mathcal{N}$-Ball semantic representation of the sentence "Mary believes herself to like Bill" shall be transformed from the $\mathcal{N}$-Ball configuration of the sentence without anaphor. In this case, the transformation shall enlarge the $\mathcal{N}$-Ball of believes-actor, so that it could partially overlap with the $\mathcal{N}$-Ball of like-actor and the $\mathcal{N}$-Ball of Mary would coincide with the $\mathcal{N}$-Ball of Jane. The coincidence conveys the meaning of herself, as illustrated in Fig. 9.9.

The $\mathcal{N}$-Ball semantic representation of the sentence "Mary believes Bill to like herself" shall be transformed from the $\mathcal{N}$-Ball configuration of the sentence "Mary believes Bill to like Jane". As 'Mary' is nearer to 'Bill' than to 'Jane' in the sentence,

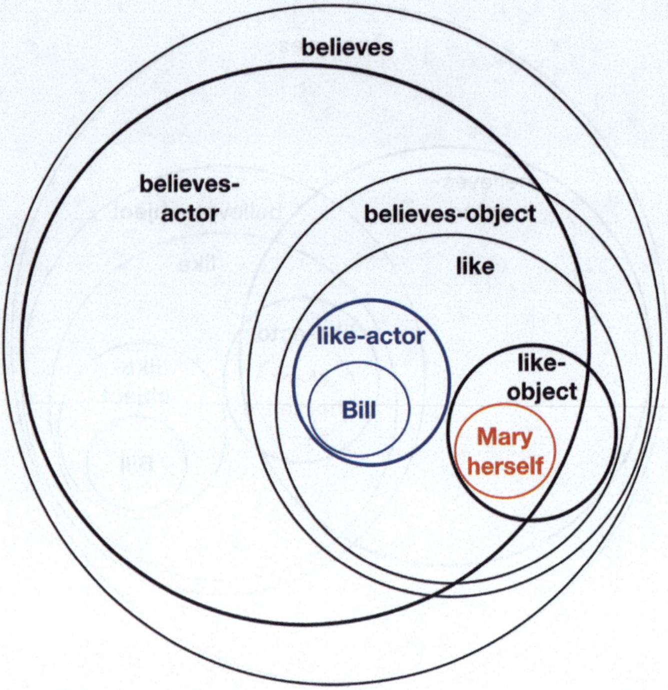

Fig. 9.10 If the $\mathcal{N}$-Ball of believes-actor is enlarged to partially overlap with the $\mathcal{N}$-Ball of like-object, it must overlap with the $\mathcal{N}$-Ball of like-actor

then $\mathcal{N}$-Ball of Mary is nearer to $\mathcal{N}$-Ball of Bill than to $\mathcal{N}$-Ball of Jane. This leads to a geometric property that if an extended $\mathcal{N}$-Ball of believes-actor overlaps with the $\mathcal{N}$-Ball of like-object overlaps, it must overlap with the $\mathcal{N}$-Ball of like-actor, as illustrated in Fig. 9.10. This introduces an unintended feature that Bill is also a believes-actor, with this innate geometric feature, we can label the sentence "Mary believes Bill to like herself" as ungrammatical. Principle A is in a way interpreted as features of geometrical operations on $\mathcal{N}$-Ball configurations.

References

Bauckhage, C., Schulz, D., & Hecker, D. (2019a). Informed Machine Learning for Industry. *ERCIM News*, *2019* (116).

Bauckhage, C., Sifa, R., & Dong, T. (2019b). Prototypes within minimum enclosing balls. In *ICANN-19* (pp. 365–376). Germany: Munich.

Cremers, A. B., Thrun, S., & Burgard, W. (1994). From AI technology research to applications. In *Proceedings of the IFIP Congress 94*. Hamburg, Germany, Elsevier Science Publisher.

de Laguna, T. (1922). Point, line and surface as sets of solids. *The Journal of Philosophy*, *19*, 449–461.

Fodor, J. A., & Pylyshyn, Z. W. (1988). Connectionism and cognitive architecture-a critical analysis. *Cognition*, *28*(1–2), 3–71.

Freidin, R. (1988). Connectionism and the study of language. *Behavioral and Brain Sciences*, *1*, 34–35.

Glasgow, J., Narayanan, N. H., & Chandrasekaran, B. (Eds.). (1995). *Diagrammatic reasoning: Cognitive and computational perspectives*. Cambridge, MA, USA: MIT Press.

Grenon, P., & Smith, B. (2004). SNAP and SPAN: Towards dynamic spatial ontology. *Spatial Cognition and Computation*, *4*(1), 69–103. Lawrence Erlbaum Associates, Inc.

Hunter, L. E. (1988). Some memory, but no mind. *Behavioral and Brain Sciences*, *1*, 37–38.

Lv, X., Hou, L., Li, J., & Liu, Z. (2018). Differentiating concepts and instances for knowledge graph embedding. In *Proceedings of the 2018 Conference on Empirical Methods in Natural Language Processing* (pp. 1971–1979). Association for Computational Linguistics.

Marcus, G. F. (2003). *The algebraic mind–integrating connectionism and cognitive science*. The MIT Press.

Minsky, M., & Papert, S. (1988). *Perceptrons*. Cambridge, MA, USA: MIT Press.

Smolensky, P. (1988a). On the proper treatment of connectionism. *Behavioral and Brain Sciences*, *1*, 1–23.

Sun, R. (2016). Implicit and explicit processes: Their relation, interaction, and competition. In L. Macchi, M. Bagassi, & R. Viale (Eds.), *Cognitive unconscious and human rationality* (p. 257). Cambridge, MA: MIT Press.

Tolman, E. C. (1948). Cognitive maps in rats and men. *The Psychological Review*, *55*(4), 189–208.

Towell, G. G., & Shavlik, J. W. (1994). Knowledge-based artificial neural networks. *Artificial Intelligence*, *70*(1–2), 119–165.

Tversky, B. (1993). Cognitive maps, cognitive collages, and spatial mental models. In A. Frank & I. Campari (Eds.), *Spatial information theory—a theoretical basis for GIS* (pp. 14–24). Springer.

Tversky, B. (1992). Distortions in cognitive maps. *Geoforum*, *23*(2), 131–138.

Tversky, B. (2019). *Mind in motion*. New York, USA: Basic Books.

Venn, J. (1880). On the diagrammatic and mechanical representation of propositions and reasonings. *The London, Edinburgh and Dublin Philosophical Magazine and Journal of Science*, *10*(58), 1–18.

von Rüden, L., Mayer, S., Garcke, J., Bauckhage, C., & Schücker, J. (2019). Informed machine learning-towards a taxonomy of explicit integration of knowledge into machine learning. CoRR arXiv:abs/1903.12394.

Whitehead, A. N. (1929). *Process and reality*. Macmillan Publishing Co., Inc.

Appendix A
Code List

Codes are pushed to github for public access.

- A python program is implemented for the experiment described in Chap. 4, and located at

  https://github.com/gnodisnait/bp94nball

- A python package is implemented for structural imposition onto word-embeddings described in Chap. 6. The source code, along with evaluation scripts, are located at

  https://github.com/gnodisnait/nball4tree

- As part of the P3ML project,[1] the nball4tree package is used to construct $\mathcal{N}$-Ball embeddings in Arabic, Albanian, Chinese, Hindi, German, Russian during the IPEC Lab "AI Language Technology". Codes are located at

  https://github.com/p3ml/ai_language_technology

- A python package is developed for Triple Classification described in Chap. 7. The source code, along with experiment results are located at

  https://github.com/gnodisnait/mushroom

[1]Funded by the Ministry of Education and Research of Germany (BMBF) under grant number 01/S17064.

T. Dong, *A Geometric Approach to the Unification of Symbolic Structures and Neural Networks*, Studies in Computational Intelligence 910,
https://doi.org/10.1007/978-3-030-56275-5

Appendix B
Sample Task for Membership Validation

We describe a sample task of Sect. 6.5.4. In Word-Net 3.0, city.n.01 has 605 members. We choose 1% of the 605 members as known city names (7 city names marked as red, following the first #), the rest as unknown city names (598 city names marked as blue, following the second #). Additionally, we randomly choose 100 entities which are neither members of the city.n.01 nor members of hypernymies of city.n.01 (marked as cyan, following the third #), and randomly choose 100 words which does not exist in the knowledge graph (marked as purple, following the forth #). The content of a task text is listed as follows. Our n-ball method correctly predicts 594 unknown cities without recognising non-city as city. This results in 100% precision and 99% recall.

city.n.01#linz.n.01 cremona.n.01 winchester.n.01 fargo.n.01 philippi.n.01 atlanta.n.01 medan.n.01#bologna.n.01 medina.n.01 toulouse.n.01 zaragoza.n.01 tokyo.n.01 katsina.n.01 fez.n.01 melbourne.n.02 manchester.n.01 montpelier.n.01 huambo.n.01 maseru.n.01 ankara.n.01 tijuana.n.01 waco.n.01 lublin.n.01 salem.n.03 lome.n.01 springfield.n.03 kigali.n.01 minneapolis.n.01 aquila.n.02 kumasi.n.01 hohhot.n.01 liege.n.03 morgantown.n.01 wuhan.n.01 charlotte.n.01 taichung.n.01 tucson.n.01 shiraz.n.01 maiduguri.n.01 dhaka.n.01 rabat.n.01 lansing.n.01 plano.n.01 ljubljana.n.01 guantanamo.n.01 geneva.n.01 leipzig.n.01 toronto.n.01 campeche.n.01 albuquerque.n.01 cherepovets.n.01 harrisburg.n.01 novosibirsk.n.01 marrakesh.n.01 tacoma.n.01 windsor.n.01 moscow.n.01 fresno.n.01 madrid.n.01 karachi.n.01 quito.n.01 aswan.n.01 namur.n.01 chelyabinsk.n.01 sfax.n.01 blantyre.n.01 kimberley.n.01 kolonia.n.01 lund.n.01 newport.n.02 sparta.n.01 dayton.n.01 gulu.n.01 lille.n.01 omsk.n.01 salem.n.02 urmia.n.02 lancaster.n.01 hartford.n.01 valencia.n.01 leeds.n.01 zomba.n.01 samarkand.n.01 mumbai.n.01 tangshan.n.01 richmond.n.01 halicarnassus.n.01 goma.n.01 cordoba.n.03 bonn.n.01 bangalore.n.01 lhasa.n.01 nagoya.n.01 astana.n.01 bursa.n.01 nineveh.n.01 katowice.n.01 amarillo.n.01 maputo.n.01 waterbury.n.01 ephesus.n.01 ottawa.n.03 montreal.n.01 volgograd.n.01 charleston.n.

T. Dong, *A Geometric Approach to the Unification of Symbolic Structures and Neural Networks*, Studies in Computational Intelligence 910,
https://doi.org/10.1007/978-3-030-56275-5

01 london.n.01 omaha.n.02 quebec.n.01 persepolis.n.01 pinsk.n.01 shenyang.n.01 toyota.n.01 versailles.n.01 milwaukee.n.01 kampala.n.01 valletta.n.01 louisville.n.01 soledad.n.01 brno.n.01 nakuru.n.01 lexington.n.02 jakarta.n.01 stockholm.n.01 riverside.n.02 worcester.n.02 apia.n.01 wheeling.n.01 lausanne.n.01 jackson.n.10 tabriz.n.01 singapore.n.01 ur.n.01 shreveport.n.01 augusta.n.02 bruges.n.01 augusta.n.01 hobart.n.01 blida.n.01 delhi.n.01 charleroi.n.01 tulsa.n.01 kuwait.n.01 mosul.n.01 kinshasa.n.01 raleigh.n.02 prague.n.01 braunschweig.n.01 caracas.n.01 malabo.n.01 binghamton.n.01 lucknow.n.01 taipei.n.01 huntington.n.04 xian.n.01 male.n.03 adelaide.n.01 kingstown.n.01 bern.n.01 scranton.n.01 portland.n.02 lincoln.n.02 tegucigalpa.n.01 venice.n.01 washington.n.01 jerusalem.n.01 taif.n.01 trento.n.01 split.n.06 trenton.n.01 vladivostok.n.01 aachen.n.01 juneau.n.01 mycenae.n.01 dobrich.n.01 suez.n.01 matamoros.n.01 soweto.n.01 jabalpur.n.01 dushanbe.n.01 leon.n.02 bolzano.n.01 nice.n.01 cali.n.01 babylon.n.01 novgorod.n.01 minsk.n.01 chemnitz.n.01 kaunas.n.01 hangzhou.n.01 pisa.n.01 bandung.n.01 st._petersburg.n.02 astrakhan.n.02 bratislava.n.01 yangon.n.01 banff.n.01 manila.n.02 arnhem.n.01 sacramento.n.01 munich.n.01 kananga.n.01 wroclaw.n.01 omdurman.n.01 brasilia.n.01 zabrze.n.01 bloemfontein.n.01 yaounde.n.01 yerevan.n.01 edmonton.n.01 delphi.n.01 topeka.n.01 paterson.n.02 khabarovsk.n.01 sapporo.n.01 pittsburgh.n.01 nancy.n.01 honolulu.n.01 columbus.n.01 guadalajara.n.01 bremen.n.01 nanning.n.01 temuco.n.01 pierre.n.01 jinja.n.01 lilongwe.n.01 antananarivo.n.01 gomorrah.n.01 boise.n.01 helena.n.01 cambridge.n.03 tepic.n.01 schenectady.n.01 limeira.n.01 victoria.n.07 solingen.n.01 casper.n.01 wurzburg.n.01 halle.n.01 bishkek.n.01 mesa.n.02 salem.n.01 gafsa.n.01 chernobyl.n.01 cheyenne.n.01 durango.n.01 samaria.n.01 flint.n.03 arequipa.n.01 seoul.n.01 annapolis.n.01 concord.n.01 fredericton.n.01 salzburg.n.01 nuremberg.n.01 nyala.n.01 cambridge.n.02 bissau.n.01 bamako.n.01 winnipeg.n.01 taegu.n.01 cleveland.n.01 brasov.n.01 abidjan.n.01 ostrava.n.01 lyon.n.01 potsdam.n.01 bangui.n.01 bakersfield.n.01 birmingham.n.01 rasht.n.01 curitiba.n.01 kabul.n.01 rosario.n.01 utrecht.n.01 independence.n.03 regina.n.01 rotterdam.n.01 timbuktu.n.01 peoria.n.01 tampere.n.01 villahermosa.n.01 nagano.n.01 olympia.n.01 kitakyushu.n.01 cologne.n.01 argos.n.01 rochester.n.01 tours.n.01 essen.n.01 dresden.n.01 kaluga.n.01 suva.n.01 daugavpils.n.01 indianapolis.n.01 beckley.n.01 troy.n.02 kandahar.n.01 wichita.n.02 pueblo.n.02 garland.n.02 hanoi.n.01 faisalabad.n.01 praia.n.01 paris.n.01 warszawa.n.01 concepcion.n.01 pergamum.n.01 lynchburg.n.01 tabuk.n.01 roseau.n.01 tartu.n.01 kishinev.n.01 providence.n.01 uppsala.n.01 brighton.n.01 thebes.n.01 arlington.n.01 reading.n.06 valencia.n.02 bogota.n.01 constantine.n.02 darwin.n.02 grenoble.n.01 apeldoorn.n.01 turin.n.01 kolkata.n.01 spokane.n.01 parkersburg.n.01 abuja.n.01 tarawa.n.01 bari.n.01 memphis.n.01 chattanooga.n.01 montevideo.n.01 baghdad.n.01 toyohashi.n.01 espoo.n.01 aberdeen.n.04 adana.n.01 cincinnati.n.01 newark.n.01 durham.n.01 rheims.n.01 bayonne.n.01 oviedo.n.01 huntsville.n.01 worcester.n.03 greensboro.n.01 utica.n.02 vienna.n.01 athens.n.01 fukuoka.n.01 winston-salem.n.01 edirne.n.01 lusaka.n.01 beijing.n.01 islamabad.n.01 tianjin.n.01 nashville.n.01 dover.n.01 austin.n.01 roanoke.n.01 mandalay.n.01 gaborone.n.01 orizaba.n.01 luxor.n.01 hamilton.n.06 assur.n.01 peshawar.n.01 brescia.n.01 mashhad.n.01 puebla.n.01 nalchik.n.01 anchorage.n.03 oaxaca.n.01 lanzhou.n.01 sydney.n.01 donetsk.n.01 merida.n.01 ariana.n.01 macon.n.01 kirkuk.

n.01 chihuahua.n.01 oxford.n.01 mazar-i-sharif.n.01 nicosia.n.01 genoa.n.01 manama.n.01 yalta.n.01 tripoli.n.02 isfahan.n.01 byzantium.n.01 vaduz.n.01 braga.n.01 camden.n.01 saratov.n.01 teheran.n.01 copenhagen.n.01 odessa.n.02 nicaea.n.01 leon.n.03 riyadh.n.01 gloucester.n.02 frankfort.n.01 allentown.n.01 johannesburg.n.01 springfield.n.01 knoxville.n.01 khartoum.n.01 caloocan.n.01 ufa.n.01 managua.n.01 birmingham.n.02 orlando.n.01 taiyuan.n.01 luoyang.n.01 bruxelles.n.01 guayaquil.n.01 osasco.n.01 qum.n.01 kandy.n.01 davenport.n.01 dodoma.n.01 verona.n.01 buffalo.n.02 billings.n.01 basseterre.n.01 almaty.n.01 albany.n.01 agra.n.01 weimar.n.01 leiden.n.01 tabora.n.01 zaria.n.01 lubeck.n.01 pyongyang.n.01 hyderabad.n.02 padua.n.01 springfield.n.02 sucre.n.02 firenze.n.01 lahore.n.01 oujda.n.01 amman.n.01 dallas.n.01 constantina.n.01 nijmegen.n.01 anaheim.n.01 bydgoszcz.n.01 medellin.n.01 libreville.n.01 amsterdam.n.01 lubbock.n.01 hyderabad.n.01 nairobi.n.01 innsbruck.n.01 giza.n.01 zurich.n.01 tbilisi.n.01 memphis.n.02 halifax.n.01 bam.n.01 decatur.n.02 granada.n.01 whitehorse.n.01 grozny.n.01 zagreb.n.01 cebu.n.01 madison.n.02 asmara.n.01 syracuse.n.01 kingston.n.03 smolensk.n.01 czestochowa.n.01 sheffield.n.01 pompeii.n.01 istanbul.n.01 nouakchott.n.01 toledo.n.02 leicester.n.02 christchurch.n.01 lima.n.01 nanchang.n.01 moron.n.02 toledo.n.01 eugene.n.02 evansville.n.01 nassau.n.01 cancun.n.01 kursk.n.01 rockford.n.01 abilene.n.01 rawalpindi.n.01 hermosillo.n.01 tamale.n.01 douala.n.01 tangier.n.01 sebastopol.n.01 kazan.n.02 philadelphia.n.01 cracow.n.01 lubumbashi.n.01 orleans.n.01 berkeley.n.02 mwanza.n.01 sana.n.01 chongqing.n.01 utica.n.01 windhoek.n.01 plovdiv.n.01 perth.n.01 omiya.n.01 saskatoon.n.01 sodom.n.02 maracay.n.01 eindhoven.n.01 coventry.n.02 papeete.n.01 vientiane.n.01 n'djamena.n.01 beaumont.n.03 nanjing.n.01 pretoria.n.01 skopje.n.01 akron.n.01 niamey.n.01 berlin.n.01 mannheim.n.01 sardis.n.01 herat.n.01 hargeisa.n.01 burlington.n.01 havana.n.01 wilmington.n.02 kathmandu.n.01 youngstown.n.01 tallahassee.n.01 bulawayo.n.01 manchester.n.02 boston.n.01 rostock.n.01 sudbury.n.01 graz.n.01 jerez.n.01 pasadena.n.01 brisbane.n.01 perm.n.01 harare.n.01 dnipropetrovsk.n.01 chennai.n.01 honiara.n.01 monterrey.n.01 provo.n.01 mysore.n.01 lodz.n.01 strasbourg.n.01 syracuse.n.02 ferrara.n.01 phoenix.n.01 accra.n.01 mexicali.n.01 ibadan.n.01 montgomery.n.03 milan.n.01 tirana.n.01 asahikawa.n.01 greenville.n.02 thebes.n.02 tashkent.n.01 calgary.n.01 putrajaya.n.01 mbeya.n.01 sofia.n.01 nablus.n.01 reno.n.01 canberra.n.01 racine.n.02 belgrade.n.01 bujumbura.n.01 herculaneum.n.01 basel.n.01 chester.n.01 rome.n.01 cordoba.n.04 clarksburg.n.01 stuttgart.n.01 northampton.n.01 port-au-prince.n.01 dortmund.n.01 brazzaville.n.01 wellington.n.02 bismarck.n.02 kharkov.n.01 charlottetown.n.01 mbabane.n.01 sarajevo.n.01 funafuti.n.01 columbia.n.03 colombo.n.01 vilnius.n.01 bucharest.n.01 kyoto.n.01 budapest.n.01 dijon.n.01 denver.n.01#stick.v.12 cartagena.n.01 malpighiaceae.n.01 genet.n.03 cowgirl.n.01 browse.n.03 aged.n.01 dialogue.n.03 heat.n.04 smoulder.n.01 operate.v.03 disarm.v.02 appearance.n.05 gunnery.n.01 reintegrate.v.01 works.n.04 defend.v.01 ebitda.n.01 feeder.n.06 associate.v.01 tytonidae.n.01 alabama.n.01 musial.n.01 liliuokalani.n.01 affirmation.n.03 cud.n.01 pass.n.08 mirror.v.01 station.n.05 dim.v.02 sled.v.01 rattan.n.03 clairvoyance.n.01 pontifex.n.01 separation.n.04 mortician.n.01 theorize.v.03 leeway.n.01 doubling.n.01 harrison.n.03 fossilize.v.02 advantage.n.03 hula.n.01 dewberry.n.02 exhibit.v.01 wilson.n.11 permutation.n.04 xhosa.n.01 withdrawal.n.02 stanley.n.02 maricopa.n.01 touch.n.01 penitentiary.n.01 palau.n.02 load.v.03 thalweg.n.01 weather.v.03 overtime.n.02 fornix.n.02 slap.n.01 foundation.n.03 delusion.n.02 consolidation.n.01 squint.v.01 heloise.n.01 canker.n.01 woodcut.n.01 expedition.n.01 chink.n.01 djanet.n.01 forum.n.01 breakage.n.02 geek.n.01 sarawak.n.01 plank.v.02 permian.n.01 brugmansia.n.01 flare.n.09 invar.n.01 cashier.v.02 stagnation.n.01 dimness.n.02 gilbert.n.02 soil.n.02 anaphora.n.01 albatross.n.02 wingback.n.02 mashriq.n.01 fund.v.05 hypothalamus.n.01 nose.v.02 rainbow.n.02 salicylate.n.01 replication.n.04 italian.n.01 fanlight.n.03 occult.v.03 subvert.v.04 discipline.n.03 fullness.n.03#boozed optima tirpitz per kilometros fsb gunships maghribi poonam centa inputs evangelion papers befehlshaber battleline acimovic hid 869,000 gogic gladu dactyls switzerland-based caratinga 41-33 bodens etlis hillview ielemia kerslake evelin gun-type assir ternana moyers nanos el-p xserve gesualdo manoharan backburner cits dola jawless vellacott preservationists niya ska lynette 88-game rumours recognitions mataka eest chromatography-mass bindman mazzoli polyesters rap/sung unecessary anura innervisions expenses govindsamy colorado-based dobiegniew transitioning god-daughter tanintharyi teliasonera trivelli 94 kg rainfall loingsech magidsohn euro365 fantagraphics ehrhardt warneke nidal guugu kiffmeyer pills pro-french 2,500-acre talks 1/3rd maariv serbs mahseer lovebugs brè batak newa calke nubians siguen unenumerated 0-100 1,707 douglas-fir

Appendix C
Sample Results of Membership Validation

See Table C.1.

Table C.1 List of membership validation. Column A is the total number of children; Column B is the number of training set; TP represents the number of true-positive predictions; FP represents the number of false-positive predictions; FN represents the number of false-negative predictions

Hypernym	A	B	Unknown hyponym	TP	FP	FN	Precision	Recall
activity.n.01	70	14	calibration.n.01	55	0	1	1.0	0.98
battle.n.01	80	16	zama.n.01	58	0	6	1.0	0.91
battle.n.01	80	32	solferino.n.02	47	0	1	1.0	0.98
be.v.01	123	13	swim.v.04	110	0	0	1.0	1.0
change.n.03	77	16	movement.n.11	61	0	0	1.0	1.0
change.v.02	249	100	relax.v.07	148	0	1	1.0	0.99
change.v.02	249	150	fall.v.26	99	0	0	1.0	1.0
city.n.01	605	7	bologna.n.01	594	0	4	1.0	0.99
city.n.01	605	545	juneau.n.01	60	0	0	1.0	1.0
communication.n.02	70	1	catch.n.05	20	0	49	1.0	0.29
composer.n.01	95	86	thomson.n.01	9	0	0	1.0	1.0
compound.n.02	110	2	methionine.n.01	72	0	36	1.0	0.67
compound.n.02	110	6	phenylalanine.n.01	91	0	13	1.0	0.88
condition.n.01	185	74	depression.n.01	111	0	0	1.0	1.0
country.n.02	167	9	georgia.n.03	146	0	12	1.0	0.92
cover.v.01	78	39	bread.v.01	39	0	0	1.0	1.0
deity.n.01	237	12	fortuna.n.01	179	0	46	1.0	0.8
deity.n.01	237	166	kartikeya.n.01	70	0	1	1.0	0.99
disease.n.01	92	37	poliomyelitis.n.01	55	0	0	1.0	1.0

(continued)

T. Dong, *A Geometric Approach to the Unification of Symbolic Structures and Neural Networks*, Studies in Computational Intelligence 910,
https://doi.org/10.1007/978-3-030-56275-5

Table C.1 (continued)

Hypernym	A	B	Unknown hyponym	TP	FP	FN	Precision	Recall
european.n.01	72	8	norwegian.n.01	63	0	1	1.0	0.98
fabric.n.01	99	60	tammy.n.01	39	0	0	1.0	1.0
fabric.n.01	99	70	worsted.n.01	29	0	0	1.0	1.0
family.n.06	174	9	rubiaceae.n.01	161	0	4	1.0	0.98
fish.n.01	104	42	snapper.n.04	62	0	0	1.0	1.0
fish.n.01	104	52	menhaden.n.01	51	0	1	1.0	0.98
fish.n.01	104	84	sheepshead.n.01	20	0	0	1.0	1.0
genus.n.02	491	148	cycas.n.01	343	0	0	1.0	1.0
group.n.01	109	2	multitude.n.03	59	0	48	1.0	0.55
group.n.01	109	99	superfamily.n.01	10	0	0	1.0	1.0
herb.n.01	111	2	legume.n.01	87	0	22	1.0	0.8
herb.n.01	111	56	oregano.n.01	54	0	1	1.0	0.98
herb.n.01	111	100	caraway.n.01	11	0	0	1.0	1.0
instrument.n.01	73	1	probe.n.02	57	0	15	1.0	0.79
instrument.n.01	73	22	spectrograph.n.01	46	0	5	1.0	0.9
island.n.01	102	72	taiwan.n.02	30	0	0	1.0	1.0
letter.n.02	70	4	y.n.02	61	0	5	1.0	0.92
letter.n.02	70	35	theta.n.01	35	0	0	1.0	1.0
letter.n.02	70	42	r.n.03	28	0	0	1.0	1.0
material.n.01	80	32	earth.n.02	47	0	1	1.0	0.98
material.n.01	80	40	pigment.n.02	40	0	0	1.0	1.0
move.v.02	87	1	turn.v.04	13	0	73	1.0	0.15
move.v.02	87	35	rock.v.02	49	0	3	1.0	0.94
move.v.03	88	80	jump.v.01	8	0	0	1.0	1.0
music.n.01	91	1	tune.n.01	88	0	2	1.0	0.98
music.n.01	91	64	rap.n.05	27	0	0	1.0	1.0
object.n.01	73	59	cave.n.01	14	0	0	1.0	1.0
person.n.01	431	345	literate.n.01	86	0	0	1.0	1.0
phenomenon.n.01	93	28	exchange.n.01	61	0	4	1.0	0.94
physicist.n.01	116	12	sakharov.n.01	94	0	10	1.0	0.9
physicist.n.01	116	105	frisch.n.01	11	0	0	1.0	1.0
plant.n.02	70	49	shrub.n.01	21	0	0	1.0	1.0
plant.n.02	70	56	endemic.n.02	14	0	0	1.0	1.0
plant.n.02	70	63	ramp.n.02	7	0	0	1.0	1.0
port.n.01	153	62	haiphong.n.01	91	0	0	1.0	1.0
process.n.06	250	50	precipitation.n.02	199	0	1	1.0	0.99
process.n.06	250	125	inversion.n.03	125	0	0	1.0	1.0
process.n.06	250	225	phenomenon.n.01	25	0	0	1.0	1.0
remove.v.01	79	32	condense.v.03	47	0	0	1.0	1.0
remove.v.01	79	40	bone.v.02	39	0	0	1.0	1.0
remove.v.01	79	56	wash.v.09	23	0	0	1.0	1.0

(continued)

Table C.1 (continued)

Hypernym	A	B	Unknown hyponym	TP	FP	FN	Precision	Recall
river.n.01	154	47	scheldt.n.01	104	0	3	1.0	0.97
river.n.01	154	62	yalu.n.01	90	0	2	1.0	0.98
shrub.n.01	76	61	blackthorn.n.01	15	0	0	1.0	1.0
state.n.01	94	66	gujarat.n.02	28	0	0	1.0	1.0
state.n.02	77	24	sale.n.04	53	0	0	1.0	1.0
statesman.n.01	99	80	bacon.n.03	19	0	0	1.0	1.0
structure.n.04	72	15	nail.n.01	56	0	1	1.0	0.98
structure.n.04	72	58	nail.n.01	14	0	0	1.0	1.0
substance.n.01	107	22	selenium.n.01	82	0	3	1.0	0.96
supply.v.01	79	48	curtain.v.01	30	0	1	1.0	0.97
supply.v.01	79	72	dado.v.01	7	0	0	1.0	1.0
town.n.01	222	12	jackson.n.08	202	0	8	1.0	0.96
town.n.01	222	156	nogales.n.01	66	0	0	1.0	1.0
travel.v.01	112	34	pursue.v.02	78	0	0	1.0	1.0
whole.n.02	99	10	septum.n.02	89	0	0	1.0	1.0
whole.n.02	99	40	annulus.n.02	59	0	0	1.0	1.0
writer.n.01	344	35	malory.n.01	306	0	3	1.0	0.99
writer.n.01	344	207	hemingway.n.01	137	0	0	1.0	1.0
writer.n.01	344	241	mitchell.n.04	103	0	0	1.0	1.0

Appendix D
The Nine Laws of Cognition

- *First Law of Cognition: There are no benefits without costs* (Tversky 2019, pp. 15, 39, 51, 53, 150, 240).
- *Second Law of Cognition: Action molds perception* (Tversky 2019, p. 18).
- *Third Law of Cognition: Feeling comes first* (Tversky 2019, p. 42).
- *Fourth Law of Cognition: The mind can override perception* (Tversky 2019, pp. 55, 64).
- *Fifth Law of Cognition: Cognition mirrors perception* (Tversky 2019, pp. 57, 73, 81).
- *Sixth Law of Cognition: Spatial thinking is the foundation of abstract thought* (Tversky 2019, pp. 72, 142).
- *Seven Law of Cognition: The mind fills in missing information* (Tversky 2019, pp.78, 244).
- *Eighth Law of Cognition: When thought overflows the mind, the mind puts it into the world* (Tversky 2019, p. 190).
- *Ninth Law of Cognition: We organize the stuff in the world the way we organize the stuff in the mind* (Tversky 2019, p. 280).

T. Dong, *A Geometric Approach to the Unification of Symbolic Structures and Neural Networks*, Studies in Computational Intelligence 910,
https://doi.org/10.1007/978-3-030-56275-5

Bibliography

Tversky, B. (2019). *Mind in motion*. New York, USA: Basic Books.

T. Dong, *A Geometric Approach to the Unification of Symbolic Structures
and Neural Networks*, Studies in Computational Intelligence 910,
https://doi.org/10.1007/978-3-030-56275-5

Index

T. Dong, *A Geometric Approach to the Unification of Symbolic Structures and Neural Networks*, Studies in Computational Intelligence 910,
https://doi.org/10.1007/978-3-030-56275-5

The manufacturer's authorised representative in the EU is Springer Nature Customer Service Centre GmbH, Europaplatz 3, 69115 Heidelberg, Germany. If you have any concerns regarding our products, please contact ProductSafety@springernature.com

Printed and bound by CPI Group (UK) Ltd, Croydon, CR0 4YY
30/09/2026
02225574-0013